Qualitative Narratives of Space and Gender

Perspectives from India

Edited by
Jagriti Gangopadhyay

B L O O M S B U R Y
NEW DELHI • LONDON • OXFORD • NEW YORK • SYDNEY

BLOOMSBURY INDIA
Bloomsbury Publishing India Pvt. Ltd
Second Floor, LSC Building No. 4, DDA Complex, Pocket C – 6 & 7,
Vasant Kunj, New Delhi, 110070

Bloomsbury, Bloomsbury Academic India, and the Diana logo are trademarks of Bloomsbury Publishing Plc.

First published in India 2026
This export edition published 2026

Bloomsbury Academic India
An imprint of Bloomsbury Publishing Plc

ISBN: HB: 978-93-61317-56-9; eBook: 978-93-61310-48-5
2 4 6 8 10 9 7 5 3 1

Typeset by Integra Software Services Pvt. Ltd.
Printed & bound in India

Bloomsbury Publishing Plc makes every effort to ensure that the paper used in the manufacture of our books is natural, recyclable, and made from wood grown in well-managed forests. Our manufacturing processes conform to the environmental regulations of the country of origin.

To find out more about our authors and books, visit www.bloomsbury.com and sign up for our newsletters.

Contents

Illustrations

Acknowledgements

This book is unique and special in many ways. Scholars from different disciplines have come together for this volume to ground their research in qualitative methodology through the domains of space and gender. Another important feature of this edited book is that it provides an opportunity for young doctoral scholars to present their work, engage with their respective fields, and expand the scope of their disciplines. In this regard, I wish to express my gratitude to Mr Chandra Sekhar, my editor from Bloomsbury India, who believed in this project and made it happen. I want to particularly thank him for shaping this edited volume and for understanding the need for multidisciplinary research in the area of research methodology.

I would also like to extend my acknowledgement to all the contributors of this edited book who trusted me with their work and adhered to all the timelines. This edited book belongs to them, and I truly hope that our contribution will be noticed by scholars who are particularly interested in qualitative methodology in India.

Introduction: Understanding the Narrative-based Approach

Jagriti Gangopadhyay

Widening the Scope of Qualitative Research

Qualitative research, as a research approach, is mostly grounded in the disciplines of sociology and anthropology. All students of sociology and anthropology are often required to take courses in research methodologies, and qualitative research forms an integral component of these courses. Some common methods that are taught to young students and have also been used by scholars in these disciplines are ethnography, in-depth interviews, focus group discussions, participant observation and case studies. Though qualitative research is grounded in these two disciplines, recently, other disciplines such as literature, history and political science are also using different forms of qualitative methods to study emerging themes in their respective fields. Expanding the scope of qualitative research, scholars of these disciplines are studying archives, mapping oral histories, using the observation method, and employing auto-ethnography to interrogate research questions in sub-disciplines such as environmental humanities, medical sociology, film studies, performing arts, and gender studies within the broader disciplines of literature, history, political science and sociology.

Tracing the evolution of qualitative research, Brinkmann et al. (2014: 1–2) argue that qualitative research should not be understood as a 'monolithic' and consensual approach to doing research in a particular manner; instead, it should be seen as a site of vibrant inquiry which is open to multiple perspectives. Highlighting the multidisciplinary nature of qualitative research, they point out that as qualitative research is mostly used to study the relationship between human lives and social phenomena, it is important to constantly channelize this multidisciplinary aspect of qualitative research for this approach to remain relevant and significant in society.

Delving deeper into this aspect of fleshing out the multidisciplinary quotient of qualitative research, this edited volume brings together the works of scholars who are interested in exploring a qualitative line of inquiry in their respective disciplines.

Exploring the Idea of 'Narratives'

Within the domain of qualitative research, this edited volume is an attempt to explore the concept of narratives as used by different scholars in their disciplines. Narrative inquiry, as a form of qualitative research methodology, has received much importance in recent times. Elaborating on the concept of narrative inquiry, Clandinin and Caine (2008: 542–3) suggest that the main aim of narrative inquiry is to comprehend various forms of experience. On the other hand, it is also increasingly being used as a tool in research methodology. Highlighting the multidimensional quotient of narrative inquiry, they note that across disciplines, narrative inquiry is used to make sense of experience through conversations, dialogues and participation in the lived realities of research participants. Adding further value to narrative inquiry is the different context of research participants and study sites which enhance narrative inquiry as a mode of research methodology. Associated with the idea of narrative inquiry is the method of narrative analysis which entails researchers to accumulate detailed description of events and shape them into stories. Though each of these stories has distinct features, nonetheless, each of these stories has a shared concern for individual experiences and the significance of each of these experiences (Polkinghorne 1995: 6–10). Although narrative inquiry is largely associated with ethnography or in-depth interviews, the use of documents, pictures, metaphors, and personal philosophies are all forms of narrative inquiry (Connelly and Clandinin 1990: 2–6). Relatedly, often scholars also use narrative constructs (Holley and Colyar 2009: 682–4), to interpret texts and archives to capture time and space and to construe a social reality of these texts and archives.

This discussion on narratives highlights their varied usage as a methodological tool, and this edited volume intends to unpack the multidisciplinary component of narratives as an emerging form of research methodology.

The Thematic Sections of the Volume

The edited volume is divided into two parts. The first part of the volume is titled 'Narratives of Space', and the second part is titled 'Narratives of Gender'. In the first part, the individual chapters use a narrative line of inquiry and explore the notion of space in different sites. Likewise, in the second part, each of the chapters uses narratives as a tool to examine questions of gender in various locations. Across the chapters, the core focus is to present the idea of narratives from a multidisciplinary lens in the Indian context. Separately, there is much scholarship on the idea of space as well as gender. In fact, both concepts have individually received significant interest from scholars and researchers across fields, and hence, both these concepts have been evaluated from a multidisciplinary lens as well. Acknowledging the works of these scholars, this edited volume brings together these two narratives with the goal of exploring how the narrative line of inquiry by scholars of space and gender, and the mutual connections that can be drawn from each of these studies, contribute to a deeper understanding.

In recent times, scholars from different fields have written and published on the idea of the field as a social and cultural construct and as an embodied space. In most of these works, the scholars have used self-reflexivity as a key tool to write about their field experiences and their position as researchers in the field. For instance, Chacko (2004: 54–61) writes about her fieldwork experience in rural India and outlines how power structures operate in the field. Reflecting on her fieldwork experiences, she notes that to maintain power balance between the researcher and the research participants, it is important to go beyond self-reflexivity and adopt practices that will ensure exchange of information, accommodate the schedule of the participants, have an open mind and recognize the contribution of the research participants. Likewise, Sinha (2017: 897–903), on her work on female sex-workers in Kolkata, highlights her many challenges as a researcher for opting to do research on a marginalized and stigmatized community in India. Explaining the issues of safety and research ethics, Sinha documents her role as a researcher navigating a complex site of study. As a researcher, Sinha found it particularly difficult to get access to her research participants and through her reflexive account she articulates how, when a researcher is introduced through an organization, there are many dynamics at play and a researcher is expected to follow the rules of the organization introducing the researcher. Biswal (2021: 422–5), in her work on village spaces, argues that 'the field' as a site for interrogation for anthropologists and sociologists should be an embodiment of interactions and critical reflections and that the field as a space should emerge from the researcher's engagements during the course of the fieldwork. However, she notes that the depiction of the field works differently for scholars from marginal backgrounds as opposed to scholars from dominant backgrounds and this needs to be acknowledged while choosing a site for conducting an ethnographic study. Thus, she points out that for democratization of disciplines such as social anthropology it is necessary to recognize the hierarchies of power and knowledge that govern all geographical spaces which eventually translate to the field for scholars of anthropology, sociology and social anthropology. Going beyond the spatial intricacies of a physical field site, Burrell (2009: 154–60) points out that in the age of the internet, it would be of value for researchers to view the field as a site of networks. Through her work on internet cafes in Ghana, she notes that cyberspace is emerging as an ethnographic space and to locate one's research in the digital space and to consider the field as a consortium of networks enables the researcher to have more flexibility and continuity with regard to the idea of the field. In fact, in recent times, with the internet boom, digital spaces have received much prominence, and several scholars are using digital ethnography as a tool to administer their research (Varis 2015: 57–60; Barratt and Maddox 2016: 704–7; Airoldi 2018: 663–6). One common point addressed by these scholars is that ethnography is more of an approach and not so much a technique and hence the idea of a field needs to be fluid and more adaptable to the constant shifts and changes in society. Building on this concept of a fluid field, the chapters in this volume, under the thematic section of narratives of space, engage with the multifaceted idea of a field. Each of the scholars in their chapters approaches their respective fields from the lens of their discipline, however, the main goal of each of these chapters is to establish that their fields are more of a narrative and have their own stories and experiences to share and not just a physical or digital site.

Similar to narratives of space, the other thematic part, narratives of gender, also attempts to explore different gender ideologies and issues through the lens of narratives, with a specific focus on gender in India. In the last two decades, gender has received much scholarly attention in India, with scholars examining the intersections between gender and migration (Mazumdar et al. 2013:54–61); highlighting the role of women in tribal communities (Xaxa 2004: 356–60); gender and climate change (Agrwal 2019: 87–121); gender and caste (Menon 2009: 94–110), and women and work (Hirway and Jose 2011: 67–82). These studies indicate the wide range of gender and for quite some time, gender has moved beyond the domain of sociology, anthropology and gender studies and also been of interest for scholars of literature, history, cultural studies and performing arts. Scholars of literature have examined the questions of gender and sexuality in a postcolonial India through close readings of different texts (Parameswaran 2002: 835–41; Mondal 2002: 920–33). Likewise, scholars of history through archival research documented gender practices in colonial India. Charu Gupta (2011: 15–30; 2020: 195–201), investigating the idea of 'sexuality' in colonial India, elaborates on the construction of Dalit female subjectivity in colonial India. Contrasting Dalit women with upper-caste women, Gupta argues that official archives during the colonial rule reveal that while upper-caste women lived behind the veil and were seen as pure women, Dalit women who performed physical labour and lived in open air were considered to be more attractive and healthier to produce more milk for their children. Gupta's works are notable and significant as she provides a detailed account of women, caste and sexuality from the archives of colonial India.

Locating the contemporary female dancer's body within the dichotomy of sexuality and Indian womanhood, Mitra (2006:70–80) explores this dilemma through her analysis of an intercultural dance-theatre project in Kolkata, India, as part of Kinaetma Theatre, UK. Trained as a dancer as an Indian classical and contemporary dancer herself, Mitra (2006: 70–80), highlights that Bharatnatyam is seen as the central classical dance form in India and this dance form expects the dancer to not have any physical contact with other dancers. During the nationalist period and in postcolonial India as well, women were mostly desexualized beings, and they were projected in line with the image of Mother India. Consequently, devadasis with their history of sexuality in the dance form were seen as a major threat to dance as an art and this led to the transformation of the female dancer who was expected to represent the chaste Indian women as per the norms of the Hindu nationalist agenda. Focusing on the nationalist image of Mother India, women and nationhood in calendar art, seminal scholar Patricia

Uberoi (1990: WS41–WS45; 2002: 193–229) argued that the image of Mother India was often used to represent a unified and secular picture of India to fulfil the goal of nation building in the Nehruvian era. In particular, Uberoi (1990: WS41–WS45) noted that all religious groups would surround the image of Mother India or Mother Goddess, and this form of a nationalist construction was deemed important in post-independence India to project an assimilation of religious groups through the larger ideology of the motherland. Building on this existing scholarship on gender, the chapters under this part use the lens of narratives to illustrate different gender practices and functions through a qualitative research approach.

The Chapters

As explained in the previous section, each of the chapters has been divided into two thematic sections. Though each of the chapters is diverse in its own right and has unique approaches, they are tied by the broader theme of narratives. Under the thematic part, 'Narratives of Space', there are four chapters. The first chapter focuses on mental health concerns among medical students studying in medical colleges in Karnataka. In this chapter, the author approaches mental health issues more as a lived experience and goes beyond the medicalized idea of mental health. Additionally, in this chapter, the in-depth narratives of the medical students are understood in relation to their medical institution, and thus, methodologically, narratives become a process to examine the embodiment of mental health in a medical space. In the second chapter, the authors explore caste as a narrative. Through their study of archives, the authors question the rigidity of caste structures with a specific focus on the Sthanikas, a small caste community of coastal Karnataka. Locating caste as a narrative in archival records, this chapter delineates the transformation of the Sthanikas as a caste group in postcolonial India. In the third chapter, the authors use narratives as a mechanism to unpack the

Tuluva identity through an interpretive reading of select Tulu films released in the last two decades. In this chapter, the films become a site of interrogation, and the authors rely on narratives as a method of analysis to understand the 'ideal Tuluva identity'. In the final chapter of this part, the authors use observational fieldwork and qualitative interactions to examine a specific site, the Saalumarada Thimmakka Tree Park, which has recently become an ambivalent site of eco-tourism and 'wilderness'. In this chapter, the authors reflect on the relationship between humans and non-human beings and highlight the role of the field site as a space of narrative that facilitates such relationships.

In the second part, 'Narratives of Gender', there are also four chapters. The first chapter presents a narrative interpretation of two texts: Samhita Arni's *The Missing Queen* and Pervin Saket's *Urmila.* In their interpretation, the authors not only engage with Arni and Saket's retelling of the marginalized female characters but also re-examine the hypermasculine and divine images of Rama and Lakshmana in the Indian epic, Ramayana. In the second chapter, the authors provide an auto-ethnographic account of a short and silent non-scripted performance titled 'Measures of Elsewhere', a performance that was developed and shaped by the authors. In this chapter, the authors locate the postcolonial-female hyphenated identity through the lens of performance. Using the narratives of self, the authors seek to examine performance as a combination of memory and space. In the third chapter, the author relies on online qualitative narratives of working mothers in STEM (Science, Technology, Engineering, and Mathematics) academia. In this chapter, the author explores the narratives of working mothers and highlights how motherhood continues to remain the dominant role fulfilment for these women as they strive to attain their career goals as well. In the final chapter of this part, the author also uses online qualitative narratives and illustrates the everyday experiences of childfree women in urban India. In this chapter, the author examines the narratives of these women as they manage stigma, interpersonal relationships and their general perceptions surrounding their conscious choice of being childfree.

The common theme across the chapters is the use of narratives to understand and describe their research as experiences. Each of the authors utilizes the concept of narrative inquiry to flesh out and places their findings more in terms of lived experiences as they aim to expand their individual fields, such as history, literature, sociology, performing arts and gender studies.

The Sustainable Development Goals

The other important theme that has been touched upon by each of the chapters is the thematic interplay of the Sustainable Development Goals (SDGs). In recent times, aligning research with the SDGs has gained much visibility as research from different countries seeks to address these goals through their work. The first chapter, which is grounded in the discipline of sociology and the sub-discipline of medical sociology, dwells on the broader theme of SDG 3: good health and well-being. Highlighting one of the most under-examined areas of research, this chapter approaches mental health issues from a medical sociology and disability studies perspective and focuses on medical students, one of the most under-studied cohorts in mental health dialogue. Relatedly, Chapters 2 and 3 can be placed under SDG 10: reduced inequalities, as each of these chapters interrogates the question of caste identities as represented in history and regional cinema. The fourth chapter, rooted in literature and the sub-discipline of environmental humanities, draws from SDG 11: sustainable cities and communities, and analyses the interlinkages between sustainable conservation and wilderness areas. The chapters under the second thematic part, 'Narratives of Gender', all address SDG 5: gender equality, as each of these chapters engages with gender constructions in different forms. As discussed, Chapter 5 questions the hypermasculine images in the Indian epic Ramayana and presents a feminist retelling of Rama and Lakshmana. In this retelling, the chapter, through close readings of two texts, offers a more relatable version of Rama and Lakshmana that would appeal to readers of postcolonial India. In the same vein, the next chapter approaches gender from the perspective of performance. On the other hand, the other two chapters, through in-depth interviews, seek to deliberate on emerging inequalities faced by women owing to their reproductive role. Thus, each of these chapters, in their own way, analyses and addresses gender identities with respect to their specific disciplines, such as literature, performing arts and sociology.

Though each of these chapters is diverse and relies on different qualitative methodologies, nonetheless, they are all connected to each other as they attempt to view their research topics as narratives and present their sites of research more as subjective experiences. Additionally, each of the chapters focuses on emerging trends of research in their respective fields as they aim to expand the scope of their disciplines in India. Finally, by bringing together these chapters, this edited volume is a conscientious effort in advancing narrative-based inquiry in the research methodology discourse.

References

Agarwal, Bina. 2019. 'The Gender and Environment Debate: Lessons from India'. In *Population and Environment*, edited by Lourdes Arizpe, M. Priscilla Stone, David Major and Priscilla Stone, 87–124. London: Routledge.

Airoldi, Massimo. 2018. 'Ethnography and the Digital Fields of Social Media'. *International Journal of Social Research Methodology* 21(6): 661–73.

Barratt, Monica J. and Amanda Maddox. 2016. 'Active Engagement with Stigmatised Communities through Digital Ethnography'. *Qualitative Research* 16(6): 701–19.

Biswal, Madhumita. 2021. 'Embodied Spatial Practices in the Field: Critical Ethnographies in Village Studies from India'. *Contemporary South Asia* 29(3): 419–32.

Brinkmann, Svend, Michael Hviid Jacobsen and Søren Kristiansen. 2014. 'Historical Overview of Qualitative Research in the Social Sciences'. In *The Oxford Handbook of Qualitative Research*, edited by Patricia Leavy, 17–42. Oxford: Oxford University Press.

Burrell, Jenna. 2009. 'Could Connectivity Replace Mobility? An Analysis of Internet Café Use Patterns in Accra, Ghana'. In *Mobile Phones: The New Talking Drums of Everyday Africa*, edited by Mirjam de Bruijn, Francis B. Nyamnjoh and Ingle Brinkman, 151–69. Leiden: Brill.

Chacko, Elizabeth. 2004. 'Positionality and Praxis: Fieldwork Experiences in Rural India'. *Singapore Journal of Tropical Geography* 25(1): 51–63.

Clandinin, D. Jean and Vera Caine. 2008. 'Narrative Inquiry'. In *The SAGE Encyclopedia of Qualitative Research Methods*, edited by Lisa M. Given, 542–5. Thousand Oaks, CA: SAGE Publications.

Connelly, F. Michael and D. Jean Clandinin. 1990. 'Stories of Experience and Narrative Inquiry'. *Educational Researcher* 19(5): 2–14.

Gupta, Charu. 2011. 'Writing Sex and Sexuality: Archives of Colonial North India'. *Journal of Women's History* 23(4): 12–35.

Gupta, Charu. 2020. 'Women, Gender and Sexuality: Changing Historiographies of Colonial India – A Bibliographical Essay'. *Studies in People's History* 7(2): 192–204.

Hirway, Indira and Sunny Jose. 2011. 'Understanding Women's Work Using Time-Use Statistics: The Case of India'. *Feminist Economics* 17(4): 67–92.

Holley, Karri A. and Julia Colyar. 2009. 'Rethinking Texts: Narrative and the Construction of Qualitative Research'. *Educational Researcher* 38(9): 680–6.

Mazumdar, Indrani, N. Neetha and Indu Agnihotri. 2013. 'Migration and Gender in India'. *Economic and Political Weekly* 48(19): 54–64.

Menon, Nivedita. 2009. 'Sexuality, Caste, Governmentality: Contests over "Gender" in India'. *Feminist Review* 91(1): 94–112.

Mitra, Ranjini. 2006. 'Living a Body Myth, Performing a Body Reality: Reclaiming the Corporeality and Sexuality of the Indian Female Dancer'. *Feminist Review* 84(1): 67–83.

Mondal, Anshuman. 2002. 'The Emblematics of Gender and Sexuality in Indian Nationalist Discourse'. *Modern Asian Studies* 36(4): 913–36.

Parameswaran, Radhika. 2002. 'Reading Fictions of Romance: Gender, Sexuality, and Nationalism in Postcolonial India'. *Journal of Communication* 52(4): 832–51.

Polkinghorne, Donald E. 1995. 'Narrative Configuration in Qualitative Analysis'. *International Journal of Qualitative Studies in Education* 8(1): 5–23.

Sinha, Shweta. 2017. 'Ethical and Safety Issues in Doing Sex Work Research: Reflections from a Field-Based Ethnographic Study in Kolkata, India'. *Qualitative Health Research* 27(6): 893–908.

Uberoi, Patricia. 1990. 'Feminine Identity and National Ethos in Indian Calendar Art'. *Economic and Political Weekly* 25(17): WS41–WS48.

Uberoi, Patricia. 2002. '"Unity in Diversity?" Dilemmas of Nationhood in Indian Calendar Art'. *Contributions to Indian Sociology* 36(1–2): 191–232.

Varis, Piia. 2015. 'Digital Ethnography'. In *The Routledge Handbook of Language and Digital Communication*, edited by Alexandra Georgakopoulou and Tereza Spilioti, 55–68. London: Routledge.

Xaxa, Virginius. 2004. 'Women and Gender in the Study of Tribes in India'. *Indian Journal of Gender Studies* 11(3): 345–67.

Section I

Narratives of Space

1

Sounds of Silence: Mental Health Narratives of Medical Students from Karnataka, India

Melvin Mathew Thomas

Introduction

Anil*, a fourth-year undergraduate medical student, requested that we continue the online interview with his camera turned off. 'Would you like this off the record?' I asked. For much of our hour-long conversation, Anil spoke with a visible smile – one that remained even as he recounted moments of distress that appeared to carry significant personal weight. He shared, with care, how mental health difficulties intersected with the demands of medical training. At another point in our conversation, Anil mentioned his aspiration to become the first doctor in his community.

Anil had earlier reflected on how mental illness might be perceived differently when attached to the identity of a future physician. He recalled hesitating to seek support, worried that such a disclosure might alter how he was viewed – not just by peers, but by those evaluating his professional suitability. Despite such reservations in the past, Anil confided in his father about his condition. He believed that his father would understand his concerns like no one else could – as his father, too, had been diagnosed with recurring depressive disorder several years ago.

To avoid possible scrutiny, Anil had relied on the medical prescription issued to his father to obtain medications for himself.

Anil was forthcoming in sharing this information during our video interview. However, his sudden request to turn off the camera took me aback. 'Anil? Would you like this off the record?' I repeated. Almost impulsively, his smile shrank. He responded with a hoarse sternness in his voice, 'What is the point of me telling you all this, then?'

In the process of narrating perceived institutional and personal prejudices, Anil chose to disclose his most vulnerable moments, illustrating the profound challenges faced by medical students in discussing mental health concerns. The interview continued, with the camera off for several minutes, before he quietly suggested switching it back on. The reason for this sequence remained unspoken, but the question he posed lingered long after the conversation ended.

Anil's account possibly suggests how certain decisions, such as relying on another's prescription, may not be reducible to ethical infractions alone. Rather, they might emerge from limited avenues for disclosure within a context shaped by anticipated moral judgement. Within academic settings, particularly in professional training programmes, implicit assumptions may exist about the kinds of minds and bodies that are seen as ideal participants: self-regulating, high-functioning, and resilient (Price 2011: 173). When these expectations appear misaligned with a student's lived experience, the decision to seek care may begin to carry a form of ethical risk. In such spaces, disability may not only be lived but also managed through acts of concealment, modification or strategic silence.

Students living with psychosocial disabilities might, therefore, find themselves navigating decisions not only around access but also around presence – how to be seen, when to speak and under what conditions vulnerability might be permitted. In turn, this has resulted in students with disabilities being recognised as constituting a particularly vulnerable demographic. Anil's narrative, in such a context, suggests that the concern is not only about being unwell, but also about how that condition might be read. Expressions of mental health difficulty may risk being interpreted as signs of inadequacy, prompting further concern about fitness, competence and sustained academic performance. These interpretations, though rarely formalized, may shape students' internal scripts, influencing how they choose to speak – or not speak – about their experiences.

This chapter draws on four interview vignettes selected from a broader ongoing doctoral study with undergraduate medical students living with one or more psychosocial disabilities. The chapter does not attempt to generalize these experiences, but rather reads them as situated responses that surface at the intersection of care, expectation and perceived institutional ethos. The focus here remains on how students make sense of mental well-being – not as a static state, but as something continuously evaluated against internal and external benchmarks. These perceptions are not framed as conclusions, but as entry points into understanding how psychosocial conditions might be lived in environments where silence can potentially function as both a shield and a strategy.

To support this reading, the study benefits from semi-structured in-depth interviews, which were pilot-tested and aimed to maximize the potential for uncovering diverse student perspectives (Jones, Torres, and Arminio 2014: 60). Ethical clearance was obtained from the Institutional Ethics Committee (IEC) of Kasturba Medical College, Manipal, Karnataka, India. Participants received a detailed Participant Information Sheet (PIS) before consenting to participate in the study.

The chapter is informed by selected vignettes from students enrolled in MBBS programmes across Karnataka. At the time of the interviews, participants were in their third year or above and had been diagnosed with one or more psychosocial disabilities. This inclusion criterion was set to ensure that participants could reflect meaningfully on their academic experiences. Each participant engaged in two online video call sessions of about an hour, with no more than two weeks between sessions.

Karnataka was chosen as the study site for two reasons. First, the state hosts over sixty medical colleges – the highest in India (National Medical Commission, 2025: 9–12). Second, a prior study among medical students in Karnataka reported a high prevalence of depression (Kumar, Jain, and Hegde, 2012: 233). Maximum variation within purposive sampling (Glaser and Strauss, 2006: 56; Merriam, 2009: 77) was employed to include a range of student experiences. Snowball sampling further extended the outreach by enabling participants to share the invitation with peers.

Transcriptions were subjected to narrative analysis, allowing the identification of themes and patterns within the students' narratives. This approach remained fluid, moving between close readings of emotionally significant interview moments and more distanced, thematic interpretations. It allowed the study to remain attentive to the nonlinear and, at times, conflicting, ways students reveal identities shaped by mental health challenges.

Participants were informed that their involvement did not entail any form of monetary or material compensation. Pseudonyms were used throughout this chapter to protect the privacy and confidentiality of the participants.

Literature Review

Mental Health among Medical Students: A Globally Recognized Concern

Over the past two decades, there has been an increased awareness of persistent mental health concerns among medical students across regions. This has produced a wealth of quantifiable studies on the matter. For example, a global meta-analysis found that depression alone accounts for 27.2 per cent of mental health concerns among medical students (Puthran et al., 2016: 452). In addition, the WHO age-wise demographic categorization (2021) indicates that medical students fall under an age bracket that is most vulnerable to mental health conditions. To elaborate, psychosocial concerns, possibly due to increased stress levels, were seen to be more prevalent among those below the age of twenty-four years.

Therefore, they represent a group whose experiences offer crucial insights into the complexity of the experience of psychosocial concerns. A geographic meta-analysis that focuses on the intersection of mental illness among medical students indicates that the prevalence of depression in the Middle East was 31.8 per cent, followed by North America at a rate of 30.3 per cent, (the rest of) Asia (30.1 per cent), South America (26.8 per cent) and Europe (20 per cent) (Puthran et al., 2016: 457). The same study also indicates that medical students were perceived to be 2.2 to 5.2 times more vulnerable to depressive symptoms than those of a similar age group found in the general population. Country-specific studies conducted across the globe, including India (Kumar, Jain, and Hegde, 2012: 223), the UK (Dahlin and Runeson, 2007: 6), Norway (Ruud et al., 2020: 36968) and Spain (Capdevila-Gaudens et al., 2021: 260359), indicate a similar pattern among this demographic.

A few of the unique concerns that have been noticed include an exhaustive and exhausting workload demanded by the programme, continuous exposure to pain and suffering in the clinical environment, an increased level of expectation to consistently excel academically and inadequate support systems to address the mental health concerns of medical students (Dyrbye and Shanafelt 2015: 140). Factors such as student vulnerability, social isolation and difficult clinical realities that may not align with the students' idealistic views of medicine prior to matriculation have also been noted among this demographic (Balboni et al. 2015: 507; Dahlin and Runeson 2007: 6).

Existing literature indicates that the presence of depressive symptoms among medical students is closely linked to lower sleep quality and increased consumption of sedative pills (Dudo et al. 2022: 109). In addition, such distress negatively impacts a health practitioner's ability to provide empathetic care to their patients (Cairns, Isham, and Zachariae 2024: 640). This could also translate to medical students' reduced participation in social activities, which could then potentially affect the formation of a strong support network. Furthermore, it is important to note that the undergraduate medical programme is considered extremely rigorous and often stressful. Therefore, it offers significant value to propose an early integration of mental health support for medical students during their period of training.

Existing scholarship, as indicated, highlights a range of factors that consistently contribute to the deterioration of mental well-being among medical students. However, these studies often overlook the subjective and nuanced experiences that shape these outcomes. In other words, the qualitative aspects of these experiences, particularly ones that emerge from the Indian context, remain underexplored.

Positioning of the Medical Student within the Discourse

Within institutional frameworks, specific groups, like undergraduate medical students, could find themselves in a transitional position. Medical students can be perceived as occupying dual roles, as both students and medical practitioners in training, navigating a liminal space between these categories. This specific demographic also faces other unique challenges. For example, a student of medicine is likely to accept the physician's yardstick of objective comprehension regarding what constitutes mental well-being. In several cases, the physician treating the student may also wear the professional hat of an educator in their shared classroom space. In addition, medical students regularly interact with traditional knowledge systems of health and illness through an inherited medical curriculum that is vast, structured, and institutionalized.

In such settings, medical students might also be expected to recover from a condition at a relatively quicker pace, considering their regular interaction with evolving medical knowledge as well as their ease and proximity to receiving treatment and care. Sociologist Arthur W. Frank sheds light on this regard, highlighting a dual obligation imposed on those living with disabilities. He argues:

> Contingency is the body's condition of being subject to forces that cannot be controlled. The infantile body is [...] its own internal needs and rhythms. Society expects nothing more, and infants are afforded some period to acquire control. When adult bodies lose control, they are expected to attempt to regain it [...] or at least to conceal the loss as effectively as possible. (Frank 2013: 31)

It is crucial to add that the episodic nature of mental health concerns (with varying degrees of severity being experienced by the individual) allows us to reimagine psychosocial disability as anything but a monolithic category. In other words, an individual may experience and comprehend a particular invisible disability differently when compared to another individual with the same condition. Or sometimes, they might even experience the severity of the disability differently at various stages, largely influenced by their immediate social surroundings. This provides us with the possibility of diverse understandings of mental health disability, with greater scrutiny placed upon how other social actors influence this experience. This chapter, therefore, aims to contribute to such a nuanced perspective, particularly when medical students choose to discuss and navigate their mental health concerns.

Theoretical Framework

Narratives (and Silence) as Rhetorical Agency

Scholars have made a compelling case to show how narratives of lived experience and illness narratives offer the possibility of disrupting dominant medicalized discourses of health and illness (Kleinman 1988: 1406). This allows one to draw from a diverse range of experiences, providing layered dimensions to abstract concepts surrounding health and well-being. Such engagements might also mitigate the risk of bracketing nuanced understandings of health and illness into restrictive, reductive and definitive categories. A narrative-centric approach also allows one to understand mental health from a subjective-experiential standpoint, enabling one to scrutinize the discourse's interaction within the larger social and moral milieu in which it operates.

In this chapter, the narratives elucidate the students' underlying anxieties and perceived potential implications for their medical profession. In other words, the disparate narratives illustrate the circumstances influencing the process of disclosing mental health concerns and shed light on how medical students with mental health concerns negotiate their identity within academic settings.

An exercise of such a kind borrows from the model of the psychiatric self-narrative, as developed by Lee and Dwyer (1995: 67), in which mental health concerns are proposed to be best understood through 'co-constructed narratives'. Early studies indicate that this approach can shape a person's experience of illness by influencing how they interpret past events, present symptoms and future expectations (Ricoeur 1984: 116).

In this chapter, the student narratives offer a space through which such contextual decisions can be reflected upon. These accounts might, therefore, open up dimensions of psychosocial conditions that remain difficult to articulate – elements that are not only concealed but, at times, challenging to reveal (Siebers 2011: 97). Furthermore, framing mental health concerns – particularly those without overt physiological indicators – through such a lens allows for the possibility of reconsidering how invisible disabilities are understood within the institutional framework. Rather than seeking clarity or consensus, these narratives gesture towards alternative articulations, ones that may remain partial, affective and shaped by their conditions of telling.

The term 'disability', in this context, draws from the social model and refers to an affective, experiential condition that directly or indirectly shapes individuals' lives in various ways. Such a reading emerges from the students' ongoing interactions with a world that is perceived as inaccessible or not designed to facilitate ease of access, especially for those with a disability.

Disability narratives, and the possibility of identifying or relating to shared experiences of such nature, may also provide moments of solidarity and relief for others (Herndl 2013: 187). Therefore, participating in such an exercise could potentially contribute to the process of providing contours to experiences that cannot always be contained in language. In other words, exercises that centre on the experiential may provide rhetorical agency and carry with them the potential to offer alternate meanings to an otherwise monolithic discourse that is centred on diagnosis and treatment.

This study draws on theoretical perspectives that view narratives and silences as forms of rhetorical agency. One possible way to understand how individuals exercise influence through language and symbolic action, often within complex social contexts, might be through an exploration of rhetorical agency. Kerschbaum (2014: 63) argues that rhetorical agency, specifically the distributed nature of agency, is about an individual's ability to make choices and how those choices are enabled or constrained within social and cultural contexts. Marilyn Cooper's work offers an ethical perspective on rhetorical agency. In her influential essay, 'Rhetorical Agency as Emergent and Enacted' (2011: 427), Cooper critiques the idea of agency – one that may be reduced to individual control or intent. Instead, she views agency as an emergent property that arises through interaction with the immediate environment that individuals inhabit.

Cooper emphasizes the potential of ethical responsibility in how individuals engage with agency, suggesting that awareness of one's impact on others is a crucial component of responsible rhetorical practice. The above frameworks also argue that rhetorical agency can transcend individual autonomy, thereby challenging an understanding of agency as something that is solely controlled by the individual. They highlight the importance of understanding agency as a collaborative and contextually shaped phenomenon. Within such an understanding, the role of silence in academia, particularly, may provide us with an additional layer of understanding.

Furthermore, disability rights scholar Margaret Price discusses the quiet, insidious exclusion experienced by those with disabilities in academic spaces (Price 2011: 6). Borrowing from Mary M. Reda's *Between Speaking and Silence: A Study of Quiet Students* (2009) and Cheryl Glenn's *Unspoken: A Rhetoric of Silence* (2004), Price argues that silence may operate as a 'powerful rhetorical strategy' and as 'forms of discourse in themselves' (Price 2011: 6). Therefore, silence, in such spaces, may operate as both a coping mechanism and a strategic form of resistance for individuals with disabilities.

The student narratives shared in this chapter, when placed against the aforementioned theoretical perspectives, also reflect how mental health narratives can be seen to pursue catalytic validity (Lather 1986: 67). To elaborate, they create spaces for participants to gain further awareness and, in turn, encourage future actions for possible social change. This can be seen to emerge from a continuous negotiation with existing dominant discourses; as a moment to redefine individual dis/abled identities and, at several times, as a device to resist perceived discrimination. Therefore, the narratives open fertile avenues to aid our understanding of how personal narratives can possibly reshape collective attitudes towards mental health concerns, deepening our grasp of the complexities of disabled identities within academic spaces.

Analysis

Narrative I: Between Noise and Isolation

Manish*, a third-year medical student, expressed regret about having to repeat a year of his medical education. During our interview, the persistent noise of trains passing near his hostel room's window became a distinct reminder of the external barriers he faces during a challenging academic programme. 'How do you manage with this noise?' I asked, to which he responded with a smile and a shrug, lighting a cigarette in defiance of the hostel rules.

Manish described the academic expectations that surfaced when he made the decision to pursue MBBS. 'I was the topper in school. Most of my friends here were, too. And now, look at us,' he said, as he set the lighter on the windowsill. 'Your confidence, your self-esteem, your self-worth – all of it takes a hit. You're paying lakhs in tuition fees, and soon, you're at a point of no return.' Manish went on to disclose that he had been diagnosed with multiple mental health issues, and admitted, 'It's scary. I won't pretend it isn't.' Seeking to comfort him, I asked, 'Isn't there anyone who understands? Someone you can talk to without fear of judgment?'

Visibly irritated, Manish was quick in his response, 'You're always judged. They call you crazy, mad, a lunatic. Even my parents were uncomfortable when I wanted to see a therapist back in twelfth grade. Now, they're okay with it, but my friends? For them, it's all about somehow coping.' He paused for a brief moment before adding, 'Help is there, but it's not something we consider. Even my friend who committed suicide never thought of it.'

Manish's words illustrated the deep-rooted stigma and isolation that students like him might have to constantly endure in a competitive academic setting. His account revealed how the norms of academia might be perceived to discourage vulnerability, possibly reinforcing a culture where silence about mental health struggles becomes habitual.

He indicated to me that students internalize the belief that showing emotional distress equates to weakness or incapacity for the academic rigour expected of them. 'What will happen even if I share my concerns? I have in the past … and that just made things worse. They said my mental illness is all a drama. They said … that I am being lazy and nothing else … It was simple, I stopped sharing.' For many like Manish, this silence could be seen as a form of self-protection, a possible survival strategy against an academic environment that may view them as less capable.

Disclosing a mental health condition, as seen in Manish's case, could also alter professional relationships with peers and faculty, often with the risk of being marginalized. This was evident in Manish's account when he narrated: 'During the initial few days of my struggle, I had friends who would regularly check on me. But then … I requested one of them for lecture notes, just a couple of days' worth of notes … They came up with excuses. They were probably worried I would score more than them in the internals … They distanced themselves from me, literally, when I lost a year.' Reflecting on this instance, Manish recounted how students are expected to continuously excel – to be, in his words, 'the goose that lays golden eggs'. Yet, any sign of weakness – a visible 'crack' in their façade – could be met with judgement and disdain.

'We're treated like a spoilt apple the moment we show any signs of struggle,' he explained, gesturing to his cigarette as though it symbolized his defiance against the constraints imposed on him by the larger academic structure. 'We keep trying to be that golden goose, but it's like Midas' touch … it becomes … a curse'.

The silence that students like Manish adopt is more than an absence of words; it could be seen as a form of survival in a space that seldom accommodates those with mental health concerns. Such a perception might have influenced his friend's reluctance to share notes. The reluctance in such an instance might also emerge to potentially uphold the institution and professional standards, ones that are measured by competitive standardized examinations. Such circumstances may lead to students who experience difficulties being perceived as unfit to remain enrolled in the medical programme.

Manish described this silence as painfully isolating, impacting his sense of self-worth and belonging. 'People only ask, '… Why did you do such a thing? Why didn't you think of us', referring to his friend's suicide. 'They say … "Why didn't you talk to us?" But in reality … they never made it safe for us to talk [about our condition]'. Manish's experience could shed some light on the perceived risks involved in disclosing mental health struggles within a competitive academic setting. Students might be caught between the need to speak of their concerns and the fear of judgement, leading to a silence that could potentially perpetuate their perceived isolation.

Manish told me that, prior to the unfortunate event of losing his friend, his college rarely discussed mental health concerns openly. He narrated, 'Support only comes after something tragic happens. It doesn't make sense.' Manish's narrative highlights his lack of support systems in this regard, both in academic settings and at home – where students might be expected to endure silently. His story further emphasizes the urgent need to foster an environment in medical education where open dialogue and support systems are prioritized.

Another train passes by.

'How do you manage with this noise?' I asked, to which he responded with a smile and a shrug, lighting a cigarette in defiance of the hostel rules.

Narrative II: The Role of Curricula in Shaping Perceptions

In India, accessing empathetic and effective mental health support is challenging for many students, largely due to the limited number of mental health professionals available. Social stigma further complicates this issue, making it difficult for students to seek help.

Meghna*, a fourth-year medical student, recounted an instance where her friend refused therapy despite it being perceived as easily accessible. 'The clinic is right there!' she narrated, visibly frustrated with her friend for not seeking professional mental health support. 'But in my group, getting help is seen as ... an extreme step. Almost a taboo', she said. A few instances earlier, she narrated that even minor visits to the psychiatric ward, outside of regular clinical rounds, would spark 'widespread gossip'.

Reflecting on this hesitation to seek help, Meghna critiqued the 'hustle culture' prevalent among medical students. 'We act as if we're "invincible" – immune to mental health [issues]', she remarked, with sarcasm and air quotes to emphasize the expectations imposed. The prevailing perception, she narrated, is that mental health breakdowns are for the 'weak', not for 'overachievers' like her fellow medical students. Such an outlook could reinforce a possible consensus that students must endure mental health challenges without necessarily acknowledging it. In Meghna's case, it highlights the possibility of academia's prioritization of intellectual rigour over the emotional well-being of the student.

When asked about the approach to mental health in her psychiatry classes, Meghna noted that instructions were heavily skewed towards acute mental illnesses, like schizophrenia, rather than more common concerns such as depression. 'We don't discuss everyday struggles ... only diseases ... and rare cases like that', she explained. She shared how psychological concerns are presented strictly as clinical conditions, with minimal focus on the psychological well-being of students themselves. The emphasis, she noted, remains on treating older demographics, such as patients with dementia, while issues affecting younger people, including the students themselves, are largely neglected within the curriculum.

It is important to note that in Indian undergraduate medical programmes, psychiatry is a sub-unit within the larger discipline of medicine, often carrying relatively less weightage of marks in standardized examinations. As a result, psychiatry is often considered a 'soft subject', as she describes, and receives minimal attention in classroom interactions.

'If you don't call it out, it never existed – best left unacknowledged ... deal with it by yourself', Meghna remarked, alluding to the expectation that personal struggles, particularly mental health issues, should remain private to uphold standards of expected professional behaviour. Her account suggests the challenging position medical students occupy with regard to the contemporary economic market. As healthcare in India increasingly aligns with high-end premium services, medical education has adapted to meet these demands, possibly prioritizing performance and productivity to satisfy the market's requirements. This includes, as noted earlier, a possible focus on an ageing population. Although contemporary research consistently suggests a high prevalence of mental health concerns among medical students, these issues are predominantly framed within the context of diagnosis and treatment. The focus, however, was suggested to fall on patient populations rather than medical trainees themselves.

Furthermore, institutional narratives often indicate the scale of medical professionals trained each year, with terms such as 'discipline' and 'rigour' positioned as essential components in shaping what some colleges describe as 'premium doctors'. These framings are frequently reiterated in official communication materials, such as college brochures, and are presented as markers of transformation and excellence. Within such environments, expressions of mental distress by students may not always find corresponding forms of recognition. Therefore, the emphasis on professional development, while important, might potentially obscure the affective dimensions of student experiences.

In such contexts, silence around mental health could be sustained not necessarily through overt denial or dismissal of the student experience, but possibly through the absence of structures that enable such concerns to be articulated without perceived risk. When well-being is framed primarily as an individual responsibility, students experiencing difficulty may interpret their condition as a deviation from expected norms. This perception might contribute to a form of hesitation, where seeking support becomes intertwined with questions of adequacy, belonging and self-worth.

Narrative III: Where Is the Justice in Accommodation?

Sabina*, a 24-year-old final-year medical student, tells me that she feels the weight of not attending class when her mental health is at its lowest. She was once allowed to miss a class because of a mental health concern. She narrated how the students accused her of being favoured.

In the past, Sabina had excelled in cultural competitions and extracurricular activities. She is a classical singer who has won several competitions across the region. 'There was a singing competition, and I was very sick. I could barely get out of bed,' she explained, 'but the cultural secretary told me, "Everyone has problems. Why can't you stop being such a baby about it?"'

My eyes quickly caught the attention of the trophies that were neatly arranged on the bookshelf behind her. 'Did you win them all?' 'I've left it behind, quite literally', she laughed. Sabina felt forced to set aside what once brought her joy, explaining, 'These are just memories now. I just need to get this degree done without thinking too much about my mental health. If I focus on that, I will spiral. It's better left untouched ... hoping the wind that carried it until here will carry it away further'.

Sabina's account possibly suggests that the presence of stigma surrounding mental health may shape how accommodations are understood and received within medical education. Such provisions are sometimes interpreted through a lens of scepticism, especially when students appear otherwise composed. Her experience points to an unspoken expectation that personal difficulty should remain unacknowledged – a form of restraint that possibly aligns with perceived professional norms.

Within highly competitive academic spaces, support structures may be framed less as enabling tools and more as distortions of fairness. This perception appears particularly pronounced in contexts where the idea of a 'level playing field' is idealized – one that implicitly equates equal treatment with equal opportunity, without accounting for differences in access or experience. In the same vein, requests for support might be quietly resisted through an institutional gaze that perceives these needs as discretionary or burdensome. Sabina's narrative, therefore, does not claim that such responses are deliberate. However, it allows us to question whether what could appear as neutral might then be shaped by broader assumptions about effort, merit and the legitimacy of asking for support.

Narrative IV

Peace as Equilibrium: Influences on Perceptions of Mental Health

During the initial moments of our interview, Manasa*, a final-year medical student, signalled her difficulty in answering my question: 'What does mental health mean to you?' This question led to a long pause, filled with a series of clicks of her tongue, silences and hesitant gestures that reflected a possible inner conflict she was facing. Manasa's attempt to articulate her understanding of mental health, as I learnt later, seemed to possibly embody the broader tension she felt between her medical training, focused on objective, clinical definitions of health, and her own, more personal understanding shaped by family and lived experiences. Her possible challenge to find the right words, punctuated by frequent pauses, could be seen as a possible reluctance to rely solely on the textbook definitions she had learned earlier in her medical programme. Instead, she seemed to have been searching for an explanation that resonated with her at a more personal, experiential level. After a few moments, she eventually arrived at an answer she described as peace.

'I think peace is in the middle; happiness comes at one end of the spectrum, and sadness comes at the other end … and … yeah, that is mental well-being for me'. Her reflection offered a way to hold opposing states together – suggesting that mental well-being, for her, might rest not in the absence of distress or the pursuit of happiness, but in the quiet space that possibly allows both to simultaneously coexist. 'Peace is the equilibrium,' she said again, gently extending her arms, as if to give shape to the balance she described. This framing of mental well-being appeared to offer a departure from the clinical definitions she had encountered in her training.

Rather than centering diagnosis or bio-medical terminologies, Manasa's account suggested that mental health, for her, might involve sustaining a form of inner steadiness. This could possibly be one that did not necessarily promise constant happiness but instead one that might have made room for movement between these two positions. In this reading, well-being could be interpreted as the presence of a manageable calm, one that could potentially hold uncertainty without collapse.

Manasa's discomfort with this approach was apparent in her hesitations, as if each pause allowed her a moment to disentangle herself from the ingrained clinical perspectives and consider a more personal understanding. For her, this process of repeatedly redefining her response may have become an act of distancing herself from what she had memorized and internalized, possibly making space to prioritize her own beliefs and value systems.

'Do your friends in medical school have the same perception of mental health as you?' I asked. Her rapid shake of the head was accompanied by an apologetic smile. 'I am slowly realizing that medical school is not that important a factor in such instances. It's always the family … or your close circle. If you come from supportive families, [then] they will understand you. Otherwise, you have the same regressive ideas about mental health, the same stigma. It is a defeated cycle.' Mansa sounded more confident in her answer this time – fewer intervals of time or stutter. As she explained, her family's support provided a foundation for her understanding of mental health that differed from the perspectives held by many of her peers.

She observed that familial attitudes towards mental health were often more compassionate and empathetic, fostering a safe space to discuss well-being without judgement. She narrated how her family provided a more meaningful understanding of mental health – one that prioritized reflection and self-acceptance.

Manasa noted that among her peers, stigma around mental health remained present – often manifesting in subtle ways, where psychological concerns were seen as less legitimate than physical ones. This possible dissonance between the quiet support she experienced at home and the caution she perceived within institutional spaces pointed to the role that social and familial environments play in shaping how students come to interpret their own concerns. She narrated:

> In medical school, if I went around sharing that I see a therapist regularly, I don't think people would be okay with it. They'd pass judgemental, offensive comments because that is where the mentality comes in. People assume that we, medicos, are very open-minded. But just look at how psychiatry is taught—it is just another subject. Professors don't take it seriously, nor do students. I know that I don't.

In her account, psychiatry often appeared marginal. It was possibly regarded as a lesser specialization, both by peers and by faculty, and approached with a degree of scepticism that seemed to extend beyond the discipline itself. For Manasa, this scepticism appeared to echo more broadly, shaping a culture in which requests for mental health support were met with doubt. Within such a context, seeking help risked being read as excess. 'A little too high maintenance', she noted, recalling how such phrases were used casually.

Manasa narrated: 'Even among my classmates ... admitting to mental health [difficulty] ... and you will be judged. I face it everyday. At first, I was reluctant [to ask for accommodations]. But one day, I asked myself, "I am already spiralling. Why am I allowing my college and classmates to make it worse?"'

For Manasa, the absence of sustained institutional support appeared to produce a condition in which students were left to manage their mental health with little formal acknowledgement of its complexity. She described a sense of isolation that did not stem solely from her condition but from the expectation that distress must present itself in consistent, measurable ways. This expectation, she suggested, proved difficult to meet. 'How can I predict if I will have any sort of energy to even get out of bed tomorrow? My mental health often cripples me … I mean physically. But then there are days when I can manage'. Her words pointed to a possible gap between lived experience and institutional reasoning – where fluctuations might not be seen as credible, unless they conformed to legible institutional categories. This mismatch, she noted, surfaced most clearly during her efforts to access even minor accommodations, which she described as 'feeling like a monumental favour that is offered'. The issue, she suggested, was not outright rejection, but a system where support seemed to depend on how clearly and consistently one could explain their needs. To elaborate, the lack of sustained, structured support for mental health prompted Manasa to reflect on the strain of having to manage alone. Without consistent empathy from peers or faculty, she noted, students like herself were often left to take on the full burden of self-accommodation – navigating not only the condition itself, but also the labour of making it understandable to others.

She recollected, 'This made me feel all empowered initially. But the emotional toll it takes to see through that day is… exhausting. That is another thing on your plate when you're struggling to barely manage!' Manasa spoke about the fatigue of constantly justifying her needs and the feeling of being viewed as 'less capable' only because she openly acknowledges her mental health condition. The repeated need to justify her experiences placed Manasa in a loop of emotional effort. She said, 'It always felt like I had to keep fighting my own battles', later reflecting that this ongoing dismissal led her to question the legitimacy of her own needs. She recalled, 'I started wondering if asking for help was a favour or if I was just being high maintenance.'

This shift suggests how institutional invalidation may contribute to a possible erosion of self-recognition; potentially interpreting basic accommodations as excessive demands.

Manasa's account possibly gestures towards a form of mental health that accommodates movement rather than stability. In holding space for uncertainty, her narrative could challenge frameworks that privilege coherence, clarity and control. Therefore, what emerges is a reflection on what it might mean to live with contradictions and to continue negotiating care within systems that might be perceived as unequipped to offer it effectively.

Conclusion

This study invites us to understand how illness narratives reveal the complex mental health challenges faced by medical students. To elaborate, through lived experiences, students convey the pressures of medical training when it is intertwined with the perceived stigma attached to mental health within academic and professional contexts. Such narratives might help us understand that conventional clinical approaches, when solely focused on objective clinical diagnoses, risk marginalizing subjective experiences; thereby possibly reinforcing existing stigma. Such a dynamic could limit students' opportunities to engage openly with their personal struggles, sustaining a possible divide between medicalized perspectives and lived realities.

Furthermore, the student narratives shared in this chapter suggest that silence carries the potential to function as a coping strategy, rather than being seen solely as a sign of passivity. Silence, in such contexts, may therefore serve as a form of resistance against institutional environments that might overlook or misunderstand vulnerability. To elaborate, a student shared, 'I don't tell anyone because I worry it will change how they see me'. Such a perception could reflect a protective measure against potential academic repercussions or perceived social judgements that emerge from stigma. This perspective of silence might also indicate the limitations within current institutional frameworks, which appear inadequate in fostering open dialogue about students' mental well-being. In addition, the curriculum's focus on severe mental illnesses is oriented primarily towards patient care, rather than including student experiences, which might contribute to feelings of alienation.

Within such a reading, the student narratives offer a space to consider silence as a form of rhetorical agency. In this light, silence may represent a strategic stance by students in what is perceived to be an unsupportive environment. This could possibly suggest that the absence of robust support structures might compel reliance on such survival strategies. The potential implications call for a reconsideration of how mental health is addressed institutionally. Prioritizing mental health within medical education may cultivate an environment where well-being is valued alongside academic and clinical performance, potentially fostering greater inclusivity and support.

Works cited/References

Balboni, Michael J., Julia Bandini, Christine Mitchell, Zachary D. Epstein-Peterson, Ada Amobi, Jonathan Cahill, Andrea C. Enzinger, John Peteet and Tracy Balboni. 2015. 'Religion, Spirituality, and the Hidden Curriculum: Medical Student and Faculty Reflections'. *Journal of Pain and Symptom Management* 50(4): 507–15.

Cairns, Paul, Aleksandra Eriksen Isham and Robert Zachariae. 2024. 'The Association between Empathy and Burnout in Medical Students: A Systematic Review and Meta-Analysis'. *BMC Medical Education* 24: 640.

Capdevila-Gaudens, Patricia, J. Miguel García-Abajo, Diego Flores-Funes, Mila García-Barbero and Joaquín García-Estañ. 2021. 'Depression, Anxiety, Burnout, and Empathy among Spanish Medical Students'. *PLOS ONE* 16(12): e0260359.

Charmaz, Kathy. 1995. 'The Body, Identity, and Self: Adapting to Impairment'. *The Sociological Quarterly* 36(4): 657–80.

Cooper, Marilyn. 2011. 'Rhetorical Agency as Emergent and Enacted'. *College Composition and Communication* 62(3): 420–9.

Dahlin, Marie E. and Bo Runeson. 2007. 'Burnout and Psychiatric Morbidity among Medical Students Entering Clinical Training: A Three-Year Prospective Questionnaire and Interview-Based Study'. *BMC Medical Education* 7: 6.

Dudo, Katarina, Emilia Ehring, Stephan Fuchs, Sabine Herget, Stefan Watzke, Susanne Unverzagt and Thomas Frese. 2022. 'The Association of Sleep Patterns and Depressive Symptoms in Medical Students: A Cross-Sectional Study'. *BMC Research Notes* 15(1): 109.

Dyrbye, Liselotte and Tait A. Shanafelt. 2015. 'A Narrative Review on Burnout Experienced by Medical Students and Residents'. *Medical Education* 50(1): 132–49.

Frank, Arthur W. 2013. *The Wounded Storyteller: Body, Illness, and Ethics*. Chicago, IL: University of Chicago Press.

Glaser, Barney G. and Anselm L. Strauss. 2006. *The Discovery of Grounded Theory: Strategies for Qualitative Research*. Chicago, IL: Routledge.

Glenn, Cheryl. 2004. *Unspoken: A Rhetoric of Silence*. Carbondale, IL: Southern Illinois University Press.

Herndl, Diane Price. 2013. 'Politics and Sympathy: Recognition and Action in Feminist Literary Disability Studies'. *Legacy* 30(1): 187.

Jones, Susan R., Vasti Torres and Jan Arminio. 2014. *Negotiating the Complexities of Qualitative Research in Higher Education*. 2nd ed. New York: Routledge.

Kerschbaum, Stephanie L. 2014. 'On Rhetorical Agency and Disclosing Disability in Academic Writing'. *Rhetoric Review* 33(1): 55–71.

Kleinman, Arthur. 1988. *The Illness Narratives: Suffering, Healing, and the Human Condition*. New York: Basic Books.

Kumar, Ganesh, Anurag Jain and Shanthala Hegde. 2012. 'Prevalence of Depression and Its Associated Factors Using Beck Depression Inventory among Students of a Medical College in Karnataka'. *Indian Journal of Psychiatry* 54(3): 223.

Lather, Patti. 1986. 'Issues of Validity in Openly Ideological Research: Between a Rock and a Soft Place'. *Interchange* 17(4): 63–84.

Lee, Richard E. and Tim Dwyer. 1995. 'Co-Constructed Narratives around Being "Sick": A Minimalist Model'. *Contemporary Family Therapy* 17(1): 65–82. https://doi.org/10.1007/BF02249305 (accessed on 2 September 2025).

Merriam, Sharan B. 2009. *Qualitative Research: A Guide to Design and Implementation*. San Francisco, CA: Jossey-Bass.

Mio, Jeffery Scott. 2020. 'Jones, James M.'. In *The Wiley Encyclopedia of Personality and Individual Differences: Clinical, Applied, and Cross-Cultural Research*, edited by Bernardo J. Carducci, Christopher S. Nave, Jeffrey S. Mio, Ronald E. Riggio, 709–12. Hoboken, NJ: John Wiley & Sons.

National Medical Commission (NMC). 2024. *Report of the National Task Force on Mental Health and Well-Being of Medical Students*. New Delhi: National Medical Commission.

National Medical Commission. 2025. 'Revised UG Seat Matrix 2024-25 on 31-03-2025'. Available at: https://www.nmc.org.in/wp-content/uploads/2025/04/Revised%20UG%20Seat%20Matrix%202024-25%20on%2031-03-2025.pdf (accessed on 2 September 2025).

Price, Margaret. 2011. *Mad at School: Rhetorics of Mental Disability and Academic Life*. Ann Arbor, MI: University of Michigan Press.

Puthran, Ranjitha, Rohan Puthran, Melvyn W. B. Zhang, Wilson W. Tam and Roger C. Ho. 2016. 'Prevalence of Depression amongst Medical Students: A Meta-Analysis'. *Medical Education* 50(4): 456–68.

Reda, Mary M. 2009. *Between Speaking and Silence: A Study on Quiet Students*. New York: SUNY Press.

Ricoeur, Paul. 1984. *Time and Narrative*, Volume 1. Translated by Kathleen McLaughlin and David Pellauer. Chicago, IL: University of Chicago Press.

Ruud, Nora, Lise Tevik Løvseth, Karin Isaksson Ro and Reidar Tyssen. 2020. 'Comparing Mental Distress and Help-Seeking among First-Year Medical Students in Norway: Results of Two Cross-Sectional Surveys 20 Years Apart'. *BMJ Open* 10(8): e036968.

Siebers, Tobin. 2011. *Disability Theory*. Ann Arbor, MI: University of Michigan Press.

United Nations. 2006. *Convention on the Rights of Persons with Disabilities*. Available at: https://www.un.org/development/desa/disabilities/convention-on-the-rights-of-persons-with-disabilities.html (accessed on 18 June 2025).

World Health Organization. 2021. 'Adolescent and Young Adult Health'. Available at: https://www.who.int/news-room/fact-sheets/detail/adolescents-health-risks-and-solutions (accessed on 2 September 2025).

2

'Sthanika to Brahmin': Narratives, Institutions and Identity in Caste Mobility

Shreyas Kolpe and Nikhil Govind

Introduction

Social mobility in India has often been described in terms of caste mobility, pointing to the central role of caste in the structure of society (Srinivas 1972: 1–2; Sanyal 1981/2022: 40–4). Mobility here refers to the ability of individual castes to effect changes to their position in the caste hierarchy over time, though often without any change to the overall structure. The fluidity of caste underlines the very nature of the caste system throughout its history, though discussion has largely focused on the construction of familiar caste identities in modernity. Social anthropologists, political scientists and of late, historians have discerned movements over the preceding two centuries. The colonial knowledge-gathering apparatus, particularly the census, is credited with asking Indians to define their identities principally in terms of caste. Caste associations began to emerge in the late nineteenth century which concerned themselves with forming a cohesive caste identity by breaking down barriers of sub-castes and uniting under a single caste name. They often petitioned census authorities to classify them higher in the *varna* scheme than was generally conceded to them. They also insisted that derogatory caste names be dropped in favour of those that properly reflected their 'true' status and origins (Rudolph and Rudolph 1960: 13–14). These processes however followed different trajectories in different parts of India. In southern and western India, caste is understood to have 'ethnicized' earlier, which refers to the formation of broad coalitions without reference to the logic of hierarchy (Jaffrelot 2000: 758).

Ethnicization of caste in South India was accompanied by the creation of 'social imaginaries' that provided an alternative to the caste model (758). For instance, the Self-Respect Movement of Periyar emphasized the Dravidian identity of lower castes who had been dominated by invading Aryans and brought under an oppressive hierarchy. It modelled its politics on an egalitarian ideal outside caste and gender. On the other hand, castes such as Yadavs in the Ganges plains overcame regional and sub-caste boundaries and then claimed to be *Kshatriyas* (763). Thus, their social imaginary preferred to claim noble descent from warrior tribes mentioned in Sanskrit epics, accepting the logic of caste and following the model of 'Sanskritization' (Srinivas 1972: 6). Over a larger historical timespan, there have also been other models for social organisation, such as sects emphasizing *bhakti* (Sanyal 1981/2022: 78–82). This demonstrates the crucial role of ideologies in the transformation of caste identities, including the support for caste mobility. It has been pointed out that 'caste was and is, to a very considerable extent, what people think of it, and how they act on these perceptions' (Bayly 1999: 7). In this manner, discourse around caste and that relating to specific castes acquired great importance, which explains the flurry of caste histories and fervent debate on status both in the past and in the present. This could also take the form of caste protest, which, instead of merely critiquing the dominant groups, questions the hierarchical precepts themselves.

The trends outlined here are generalizations from macro-studies that treat South Asia as a whole. While these are based on studies of individual castes and local phenomena, they are likely to miss details and subtleties due to the need to find suitable comparisons. It is possible to juxtapose against this a micro-study, such as the investigation of a numerically small caste – the Sthanikas, confined to a few districts in coastal Karnataka. While macro-studies provide perspectives, micro-studies yield insights (Srinivas 1972: 2). This chapter undertakes a study of the Sthanikas since the inauguration of colonial modernity, namely in the nineteenth and twentieth centuries, with a focus on the role of narratives, institutions, and identity in enabling upward caste mobility. This includes a detailed discussion of a peculiar caste history, which has often been noted in passing as a curious development in studies on caste and historiography in India. However, where there is little published material, as is the case with the Sthanikas, it is a powerful indicator of changes in social status. We can read the effects on discourse of the reformulation of caste identity as a religious and sectarian identity through affiliation with the Sringeri *matha*, which otherwise provides little by means of symbolic representation in comparison to some other religious movements in the history of South India. Through a study of archives—those of the state and those gathered from the community – this chapter traces discourse spanning the transformation of Sthanikas from unchanging objects of ethnography to caste subjects of history.

Caste Discourse: Thoughts on Methodology

Caste is essentially a regional matrix of *jatis,* that is, hereditary endogamous groups, and a study of a single caste needs to be anchored in its regional context. The erstwhile district of South Canara is linguistically and culturally diverse, forming the border between Karnataka and Kerala, and hosting many migrant communities. Most of the 'native' Tulu-speaking castes are marked by a custom of matrilineal inheritance and the worship of spirits, referred to as *Daivaradhane*. The most numerous of these castes are the Bunts, Billavas and Mogaveeras. Brahmins also form a large section of the population, more so than in other districts of Karnataka. They are associated with the many temples dedicated to *puranic* gods and sectarian *mathas* or monasteries. Historically, they have also been large landholders alongside the Bunts. There are smaller pockets of Jains, Christians and Muslims, as well as the Konkani-speaking Goud Saraswat Brahmins. In this milieu, the Sthanikas are a small caste who were traditionally classified as temple servants, though they claim to be Brahmins. It has been observed that while the two ends of the caste hierarchy were relatively stable, there was considerable ambiguity about the ordering of castes in the middle (Srinivas 1972: 9). However, the case of the Sthanikas shows that even the category of the Brahmin was unstable in a regional and local setting, and that various strategies needed to be employed to achieve recognition. The existence of a hierarchy within the Brahmins of a given region has been noted. In Bengal, the 'clean' Brahmins were ranked above the Patit or fallen Brahmins, the Acharya Brahmins who made a living through astrology, the Agradani Brahmins who accepted prohibited food and gifts during funeral rites, and the Varna Brahmins who ministered to the 'unclean' Sudra castes (Sanyal 1981/2022: 45). The Sthanikas achieved their mobility over the course of the twentieth century through the deeds of individuals as well as institutions. Integral to this process was discourse enveloping the Sthanikas, which they attempted to re-articulate. Today, they are relatively well off in terms of economic status and have also been accepted as Brahmins for purposes of inter-marriage and commensality.

This chapter aims to contribute to the literature on caste, especially Brahminism in Karnataka, by juxtaposing two sets of narratives – those derived from traditional accounts and recorded by colonial ethnography, and select portions from a tract that can be treated as a Sthanika caste history. The objective is to map the evolution of caste discourse through its preoccupation with history and narratives about place, and, in this manner, shed light on the mobility achieved by the caste. This approach is necessitated by a relative dearth of archives concerning a numerically small and politically insignificant community, and a region with few social histories. However, a precocious instinct to record in print has preserved such a trove of material as a caste census, essays, speeches, address books, biographies and booklets. Among the last of these, there are several caste histories authored by different individuals and from varying perspectives. This assortment of material, including caste literature, may be termed a 'community archive'. A definitive sectarian turn can be witnessed in the writings of the Sthanika activist M.S. Achyuta Sharma. This chapter chooses a polemical history of Udupi authored by him to probe the formation of a Sthanika identity through discourse on religion and sect, specifically the role of the *matha* institution. The 'material' advances of the caste are read in the light of discursive shifts, corroborated through a close reading of the community archives.

Region and Caste in Tradition

In Brahminical narratives, India's west coast was reclaimed by the legendary Parashurama from the sea (Rao 2005: 12).[1] In this light, coastal Karnataka is seen as a land endowed by divinity with numerous places of pilgrimage. Udupi is the first in a list of seven holy places enumerated in a popular verse (Rao 2002: 61).[2] It is also the centre of the Madhva sect founded by the philosopher of *dvaita* or dualism, Madhvacharya, in the thirteenth century CE. The town is centred around the famous Krishna temple and several *mathas* of that sect. There are other ancient temples to various gods throughout the district that are associated with the Brahmins who serve as priests. In the everyday conduct of temple activities, there are numerous functions that have historically been assigned to different castes. The highly differentiated nature of temple service manifested in activities ranked from high to low, such as assisting the priests, preparing the articles of worship, picking flowers, maintaining the granary, keeping accounts, playing musical instruments and cleaning, among other tasks. The Sthanikas are one such caste that define themselves through their association with temples. Their roots are in the villages of this region, though they have clustered in urban centres such as Mangalore and Udupi, and have migrated to other towns of Karnataka and Maharashtra. The caste is estimated to number less than 50,000 people (Devananda Bhat, pers. comm., April 2023). Sthanikas identify themselves as Tulu Brahmins, of the Smarta sect affiliated to the Sringeri *matha* and as having been managers or administrators of the temple in the past. However, this self-identification as Brahmins was historically denied to them, marked by a bar on priesthood, certain types of religious learning, and in being consigned to perform tasks in the temple considered lowly. This is corroborated by memory and certain elements of Sthanika discourse on caste which claim a recent period of marginalization and humiliation at the site of the temple.

The Brahmins are divided into various sub-castes, including the more populous Tulu Brahmins and smaller groups of Marathi and Konkani speakers, who are considered to be later migrants (Sturrock 1894: 144). The former are identified as Kota, Koteshvara, Kandavara, Shivalli and Havyaka Brahmins. The first four are names of *agraharas* or Brahmin settlements, which are, in fact, the largest of the thirty-two settlements attributed to King Mayuravarma, who supposedly brought them from a place called Ahicchatra in the north (Rao 2005: 12). Historically speaking, numerous inscriptions containing land grants testify to the importance of these villages as agrarian centres and places of religious significance. The Shivalli village seems to have been the largest of these, containing several smaller hamlets by the late medieval era. The modern town of Udupi developed from a small locality in this larger village. Shivalli was already a prestigious religious centre by the eighth century (Adiga 2001: 64).[3] The name, literally 'Shiva's village' in Kannada, is reasoned to have emerged from the Ananteshwara temple dedicated to Shiva, considered to have preceded the more famous Krishna *matha*. The latter made Udupi into the centre of a Vaishnava devotional cult throughout South India. Most Shivalli Brahmins belong to the Madhva sect while a small minority are Smartas (Sturrock 1894: 146). Thus, whenever Brahmins are mentioned in this chapter, it usually refers to this sub-caste and to their religious practices. Temple priests, house priests, ritual experts, as well as religious scholars have generally come from the ranks of Shivalli Brahmins. It is also against this dominant group, as we shall see, that Sthanika polemics were directed.

The Sthanikas in Colonial Ethnography

In 1799, Tipu Sultan was defeated in the Fourth Anglo-Mysore War, and the tract of coastal Karnataka entered Madras Presidency as the territory of Canara (Bhat 1998: 4). Thus, a neat line can be drawn in the history of the region, separating modernity, colonial rule and the nineteenth century from the pre-modern or early modern past. In 1800, a Scottish surgeon and botanist named Francis Buchanan was deputed to undertake a survey of the region along with the newly added territories of Mysore and Malabar. The report of this journey would be followed by those of Colin Mackenzie, revenue settlement surveys and the decennial census beginning in 1871. The information gathered in these surveys made its way into various reports and gazettes, such as the South Canara 'district manual' of John Sturrock published in 1894. Interest in identifying, describing and classifying the inhabitants would persist into the twentieth century, such as in the mammoth project by Edgar Thurston to produce the *Castes and Tribes of Southern India* in eight volumes in 1909. Such systematic knowledge gathering has been understood to reveal the intricate relationship between knowledge and power under colonialism. Bernard Cohn (1996) finds the use of several 'investigative modalities' to gather facts, namely historiography, observation and travel, survey, enumeration, museology and surveillance (5). A result of this was the 'objectification' of India where its society, history and culture were packaged as a 'thing' and coded 'in ways that rendered it increasingly available for colonization' (Dirks 1996: xv).

The travelogue of Buchanan is brimming with detail, and the similarities to ethnographic accounts produced a century later are striking. An interesting table is the 'Kaneh Shumareh, or statement of Casts [*sic*], Men, Boys, Women, and Girls,' with 122 'Casts or

Trades' that includes even the most marginal groups, with a handful of people counted under it (Buchanan 1807: 5). Each group is supplemented by a brief occupational description. In the list starting with the 'Brahmans', the 'Stanicas [*sic*]' are listed as the sixth entry, separate from the Brahmins and described as 'employed in low offices at heathen temples' (5). As Buchanan journeys through the country, he describes various castes as part of the narrative, including the 'Tuluva Brahmins', who 'consider themselves as the proper lords of the country' (16). It will be useful to read his note in full since it is the earliest colonial account of the Sthanikas and since the ethnographic details would be repeated in subsequent surveys.

> In the temples of *Tulava* [*sic*] there prevails a very singular custom which has given origin to a cast named *Moylar*. Any woman of the four pure casts, *Brahman*, *Kshatri*, *Vaisya*, or *Sudra*, who is tired of her husband, or who (being a widow, and consequently incapable of marriage,) is tired of a life of celibacy, goes to a temple, and eats some of the rice that is offered to the idol. She is then taken before the officers of government, who assemble some people of her cast to inquire into the cause of her resolution; and, if she be of the *Brahman* cast, to give her an option, of living either in the temple or out of its precincts. If she choose the former, she gets a daily allowance of rice, and annually a piece of cloth. She must sweep the temple, fan the idol with a *Tibet* cow's tail (*Bos gruiens*), and confine her amours to the *Brahmans*. In fact, she generally becomes a concubine to some officer of revenue, who gives her a trifle in addition to her public allowance, and who will flog her severely if she grant favours to any other person. The male children of these women are called *Moylar*, but are fond of assuming the title of *Stanika*, and wear the *Brahmanical* thread. As many of them as can procure employment live about the temples, sweep the areas, sprinkle them with an infusion of cow-dung, carry flambeaus before the gods, and perform other similar low offices. The others are reduced to betake themselves to agriculture, or some honest employment. The daughters are partly brought up to live like their mothers, and the remainder are given in marriage to the *Stanikas*

> The *Moylar* differ much in their customs, each endeavouring to follow those of the cast from which his mother derived her origin. Thus the descendants of a *Brahmany* prostitute wear the thread, eat no animal food, drink no spirituous liquors, and make marks on their faces and bodies similar to those which are used by the sacred cast. They are not, however, permitted to read the *Vedas*, nor the eighteen *Puranas*. Indeed but very of them learn to keep accompts [*sic*], or to read songs written in the vulgar language. Contrary to the custom of the *Brahmans*, a widow is permitted to marry. They burn the dead, and believe in the transmigration of souls, but seem to have very crude notions on this subject. They are, indeed, very ignorant of the doctrine of the *Brahmans*, who utterly despise them, and will not act as their *Gurus* to give them *Upadesa*. They will attend, however, at the ceremonies of the *Moylar*, and read the services proper on the occasion, and will accept from them both *Dhana* and *Dharma*. (65–6)

There are many interesting details in this source, but a few salient ones are that the Sthanikas were connected to the temple, were considered to be of illegitimate parentage, and that the Brahmins held them in contempt and denied them religious knowledge. In addition to their duties in the temple, some had learned to be accountants. Finally, the significance of this account is its very early date, which attests to the fact that the Sthanikas were identified distinctly already by the dawn of the nineteenth century and that their social stratification must have been long in the making.

The district manual of 1894 and the *Castes and Tribes* volumes of 1909 provide some more information about the castes that we are concerned with. In the district manual, John Sturrock lists the castes in approximate social precedence, starting with Brahmins and their twelve subdivisions (Sturrock 1894: 144–54). Immediately following the Brahmins but before the group called Traders, are listed the class called 'Temple Servants' containing four castes – the 'Stanikas', Devadigas, Moyilis or Moylars, and Sappaligs (154–5). The very first line in the note on the Sthanikas records the narrative of their origin, which is equated with 'Manu's *golaka*'.[4] They are recorded as claiming to be 'Siva Brahmins, forcibly dispossessed of authority by the Madhvas'. The Sthanikas claimed that their caste name 'indicates their profession as managers of temples, with the title of Deva Stanika'. Sturrock, however, notes that the duties of the Sthanikas are 'clearly those of the temple servants, namely collecting flowers, sweeping the interiors of temples, looking after the lamps, cleaning the temple vessels, ringing bells and the like'. He states that their claim is not 'generally conceded', perhaps by the Brahmins, but we can infer not by Sturrock himself. Many are said to be landowners and farmers, and a few educated men had become priests (154). Turning to the following three castes, they are all said to be related groups known by different names. The Devadigas are said to be a class of servants, mostly musicians in the temples. 'They are also known as Moyilis', of whom it is said that they are 'descendants of children of women attached to the temples', and that 'their ranks are even now swelled in this manner'. Their duties are said to be similar to those of the Sthanikas (155). It is here that the identification of the Moylars becomes possible, by being linked as a sub-division belonging to the larger Devadiga caste. Regarding the Sthanikas, Thurston quotes Sturrock verbatim except for a few parts where more up-to-date knowledge is offered. The most significant addition is the claim made by the Sthanikas on the 'famous temple of Subramanya' which 'is said to have been in charge of the Sthanikas, till it was wrested from them by the Shivalli Brahmans' (Thurston 1909c: 403). Some privileges at the temple are said to have been accorded to the Sthanikas in the past.

The Sect as Identity: Smartas and Madhvas

The significance of Udupi has already been introduced in this study. Though the Brahmins of coastal Karnataka were dispersed throughout the countryside, the temple town of Udupi was the seat of learning and centre of high culture. The historical core is marked out as a religious space – a square consisting of the Ananteshwara and Chandramoulishwara temples, and the Krishna *matha*.[5] Placed around the square are eight *mathas,* collectively called the *ashta-mathas,* whose *sannyasis* or ascetic pontiffs are properly the priests of Krishna at his temple, taking turns every two years. The founders of these *mathas* are thought to have been specially chosen by Madhvacharya to worship the Krishna idol of Udupi, symbolized by receiving bronze idols from his own hand of various Vaishnava deities (Rao 2002: 54–5).[6] These are worshipped separately in each of the *ashta-mathas*. Udupi is thus marked out as a distinctly sectarian space, drawing devotees of Krishna. This specific deity, in its embodied stone idol and unique iconography, is the centre of attention to the Madhvas since it is believed to have been installed and worshipped by Madhvacharya himself.

Smartas are another sect of Brahmins in southern India who identify as followers of *Advaita Vedanta,* as preached by Adi Shankaracharya (Nanjundayya and Iyer 1928: 307).[7] They are also a community marked by affiliation to *mathas* and their *gurus*, though perhaps this is less central to their identity.[8] Some of the Brahmin sub-castes of the region are identified as wholly belonging to the Smarta sect, while the Goud Saraswat Brahmins contain both Smarta and Madhva families (Thurston 1909b: 389). A sense of rivalry between the Smartas and Madhvas can be sensed, though this is often benign since marriage ties are permissible across the boundary (Nanjundayya and Iyer 1928: 534).[9] Smartas may be described as polytheists or pluralists, allowing for the worship of different deities interpreted in the light of *Advaita* as manifestations of the same *Brahman* – the ultimate reality of the Upanishads. Further, *Brahman* is identical to the individual soul – *Atman* (Lorenzen 1983: 159). On the other hand, Madhvacharya identified *Brahman* with Vishnu and held that the individual soul – *Jivatman*– is eternally separate. He founded a Vaishnava order, though other deities, including Shiva, are recognized as subordinate deities (Nanjundayya and Iyer 1928: 536; Stoker 2016: 2). Madhva philosophers identified Adi Shankaracharya's

Vedanta as their 'philosophical nemesis' (Stoker 2016: 21). However, the Smarta-Madhva division has also been cast in the mould of a Shaiva-Vaishnava devotional conflict which is again linked to a larger history of religion in India. The identification of the Smartas with Shaivism is likely due to the much wider spread of devotion to Shiva in South India. Valerie Stoker (2016) considers the role of Vyasatirtha at Vijayanagara in the early sixteenth century to be crucial in garnering imperial support for the Madhva sect and in expanding its reach. An analysis of his polemics and hagiographies reveals not only philosophical debates with *Advaitins* but also social competition with Smartas who were influential at the court (3, 7). Vyasatirtha's imperial patronage spread Madhva institutions and doctrine to new territories such as the Tamil- and Telugu-speaking regions while also expanding the economic and social networks of the empire (17). Further, going over the problematic of defining such terms as 'Hinduism' and a sect within it, she advances the term *matha* itself to convey the meaning of sect, at least for the Vijayanagara period, since the term transcends the architectural space to refer to 'conceptual entities', namely 'shared practices and ideals' (9).

Articulating Caste History as Concealed History

Caste histories of the Sthanikas began to be composed in the twentieth century which combined oral narratives with insights from the developing field of Indian historiography. These writings are part of a vast corpus of vernacular histories that have been called 'old social histories' (Chatterjee 2008: 16). These are tracts that provide socio-historical evidence for 'the many contentious battles over sectarian, caste, linguistic, ethnic, and other claims' (16). It is significant that there is little evidence for Sthanika caste histories belonging to a long lineage of traditional accounts. Instead, we have the impression of a distinctively modern genre that was created in response to the information accumulated by the historiographic and survey modalities of the colonial state.[10] The relationship of caste histories to academic histories has also been pointed out, as 'never entirely abjuring the practices of academic history but always running at a tangent to it' (17). Their writers paid close attention to academic papers and journals, borrowing liberally from their contents though not always paying attention to their methods. This section presents one such caste history, marking a transition in discussion from late-nineteenth-century ethnographies to mid-twentieth-century polemics. In 1969, M.S. Achyuta Sharma, a Sthanika caste activist, published a tract in Kannada with the title 'the true picture of Udupi and its history'.[11] Sharma's book of more than a 100 pages is an idiosyncratic text, appearing at first glance to be a congeries containing a narrative of 'history' – or rather a 'hidden history' followed by extracts from certain court cases, district manuals and *Castes and Tribes*, an academic paper on the Sthanikas and reproductions of a public debate between *Advaita* and *Dvaita*. We shall examine selected portions of the core 'historical' section in this chapter.

'Having heard that Udupi was now famous in relation to Krishna, I wished to have a look for myself', begins Sharma caustically in the introduction (Sharma 1969: 1–2). After his visit he is left feeling strange since one who is called 'lord of the earth' is cooped up in a small room without ventilation or a grand *gopura*. That the two Shiva temples of 'Shiva's village' had not a single Shaiva or Smarta Brahmin in sight but was filled with Madhvas and that Shiva was worshipped improperly filled him with doubt about the truth of the place.[12] He starts the main section by describing Udupi as an ancient *shaiva-kshetra*. Around it there are eight temples to Durga and four temples to Subrahmanya. These are said to be undoubtedly related to Shiva, and he claims that *all* of these were under the management and worship of 'Sthanika Shiva-brahmins' before the 'entry' of Madhvas (3). The two temples to Shiva are surrounded by four large courtyards, of which the northeastern corner is 'encroached' by the Krishna temple. The eight *mathas* are described as sentries blocking the Shiva temples from all sides and preventing others from entering (4). The tightly packed shopfronts on the outer temple walls makes it impossible to know of their existence. If anyone asks about them, the Madhvas are said to prevent their visit by saying that there was no need to go to any other temple after having had *darshana* of Krishna (5). The Ananteshwara temple is described as having a wide-open courtyard, with a stone sanctum of colossal pillars topped by a roof, also of stone (6). A *shiva-linga* of enormous proportions is enshrined within. After the Madhvas entered, they tried to alter the deity into 'Anantasana'.[13] The Vaishnava priests of today are said to be ignorant of *shaiva-agamas* and instead worship 'their' *saligrama* stone kept in front of the *linga*.[14] A stone *nandi* facing the deity is said to have been removed and deposited in a temple far away. The Chandramoulishwara temple is described in a similar vein. Finally, he moves to the Krishna temple. It is said to be built not on independent land but on rented premises of the Ananteshwara temple. The rent is paid every year in the form of Krishna offering his own chariot to the landlord during that deity's annual festival (11). One enters the temple from the south and enters a congested courtyard from the east where one peeps on Krishna through a window of ventilation holes on the western wall of the shrine.

> One feels sad, strange, and surprised to find such a famous Krishna a prisoner in a congested, windless room without a *gopura* or on his own land. The seers who claim to be his foremost devotees live in palatial houses but don't keep their deity in a grand temple. By worshipping him furtively they seem to be teaching *bhedavada* by example.[15] (12)

Sharma then claims that it has 'become known' that behind the Krishna-with-a-churning-rod is another idol of *Venugopala-Krishna* facing the east (13). This idol of Krishna-playing-the-flute does not receive worship, but all endowments to the temple are said to come to him. He then claims that it is the same icon as the one installed by a disciple of Adi Shankaracharya called Hastamalaka-acharya mentioned in the hagiography *Shankaravijaya* by Anandagiri (13). Next follows the astonishing claim that there is no evidence for Madhvacharya's activities in Udupi. Vadiraja-swami, a pontiff of that tradition in the sixteenth to seventeenth century who was well versed in black magic, gained favour with the people of the village and its leaders and forcefully snatched the Shiva temples, those of Venugopala and Subrahmanya, and the temple tank from the 'Sthanika Shiva-brahmins' (14). To make the Venugopala temple his own, Vadiraja-swami tried procuring some idol of Krishna and when he failed to find one, took an idol of Durga and presented it as that of Krishna with his churning rod (15). When unbearable troubles visited him as he tried to remove the Venugopala icon, he let it remain but concealed it. In the ensuing conflict, he was supported by the Jains, Bunts and Mogaveeras, while the Sthanikas were assisted by the Billavas and Untouchables. The quarters of the Sthanikas around the temple square were taken over and given to disciples from eight villages who had rendered help, to build their *mathas*. The Jains, Bunts and Billavas were made into Madhva Brahmins and thus Udupi swelled with their numbers (16). How did Vadiraja-swami gain this opportunity? Sharma lays the blame on the state of anarchy in the region due to the fall of Vijayanagara and the military campaigns of the Keladi Nayakas when the Sthanikas were left with no support. Dewan Poornaiah of Mysore, a Madhva Brahmin, is supposed to have helped his co-religionists' rise in the nineteenth century. The Sthanikas who supported their own Dewan Lakshminarayana in a mutiny fell from favour with the British.[16] It was only after 1845 that most temples went into the hands of the Shivallis (20–1).

This subversive history by Sharma succinctly captures the multiple leaps made in formulating a Sthanika identity since the late nineteenth century. It represents the major strand of counter-narratives opposed to deprecating depictions in traditional accounts that were fortified by inscription in colonial ethnography. What were the conditions that made this articulation possible? Underlying it are advancements in material prospects and in discourse, each upholding and enabling the other. This presentation of a polemic allows us to trace an institutional history of the Sthanikas in the twentieth century that allowed for a different expression of their identity.

Organizing the Sthanikas in the Twentieth Century

In the nineteenth century, the records show the Sthanikas as objects of ethnography in which they were defined consistently as a class of low servants in the temples. The Sthanikas themselves are largely silent. However, from the early twentieth century, their voices can be heard. There is definite record of the Sthanikas organizing themselves in the form of a caste association as well as a religious community associated with traditional institutions of the temple and *matha*. The emergence of caste associations in the late nineteenth and early twentieth centuries across India has been studied, especially of those claiming to represent numerically large castes and their political aspirations (Rudolph and Rudolph 1960: 5–8; Arnold, Jeffrey and Manor 1976: 353–71). There are, however, a few features in common in their origins, such as the appearance of new elites and the recognition of shared interests and grievances. In 1908, the Shri Subrahmanya Sabha was established after a few Sthanika caste leaders had met twice, resolving to set up a permanent association.[17] They had assembled on 12 July 1908 to debate a 'matter of religion' (Shri Subrahmanya Sabha 2008: 8). In the discussions they agreed to start an organization that would encourage education, bring uniformity in the observances of the community, decide on internal caste matters and contribute to the general uplift of the caste. It was resolved to assemble again after inviting representatives from across the district. On 30 and 31 December of that year, they met to establish the Sabha under the presidency of Kumbhat Krishnayya and elected its managing committee. Regional representatives were nominated to propagate news of the organization, all along the length of the

South Canara district, in Sringeri over the border in Shimoga district and Madikeri in the Coorg province. On 26 and 27 December 1909, the Sabha held its first general body meeting in the house of Nattoji Vishnu Rao in Puttur with monetary contributions from sixty-three members. In the report from that meeting, the organization counted 124 members in all.[18] Caste associations have been described as having characteristics of both the 'natural' association of caste which is ascriptive and a voluntary association that consists of members, meetings and agendas (Rudolph and Rudolph 1960: 8). They also have a potential membership and an actual membership, with the latter presuming to speak on behalf of the entire caste.

There is little information about the caste's prior organization, though it is likely that it was based on local networks of marriage ties constrained by geography. The founder-president, Krishnayya, is credited with undertaking decades of preparatory work in bringing together elders and enthusiasts.[19] Starting from his base near Subrahmanya, he travelled as far as Udupi and Madikeri in either direction on foot, at a time when travel and communication facilities were poor (Rao 2012: 2). By 1909, he had prepared a detailed caste census of over 600 families, recording their names, *gotra*, age, occupation and property. The early activities of the Sabha were aided by the presence of a small number of families that owned land, whose members had obtained education and were employed in the towns, many as doctors of Indian medicine. It has been argued that the leadership of caste associations was different from the traditional caste authorities – the leaders were young men who were misfits, assessed by the occupations usually ascribed to the caste (Arnold, Jeffrey, and Manor 1976: 372). There are several indicators of substantial social differentiation within the Sthanikas by the beginning of the twentieth century. Ethnographers had recorded that some owned land and others kept accounts. It is claimed that many Sthanikas entered the colonial village bureaucracy as *shanbhogues* or village accountants and as *potails* or village headmen. The records of the Sabha from its early years inform us of its office holders at inception and subsequent annual sessions. Apart from a few well-known figures who are supplied with biographical sketches, it is difficult to infer their social status and occupations. In 1942, it was registered under the Societies Registration Act, 1860 and adopted a new constitution (Shri Subrahmanya Sabha 2008: 12). The memorandum of association lists the directors along with their occupations. Among the seven directors are two landholders, an astrologer, a supervisor of cooperative societies, a medical officer at the municipal dispensary, a medical practitioner and the head clerk to the district education officer (Shri Subrahmanya Sabha 1942: 3). There appears to be a mix of hereditary rural landholding and urban literate occupations, especially in government service.

Apart from representing the caste to the public, the association also takes up measures for internal reform. The Sthanikas might have felt an acute lack of religious training and faced a shortage of priests to perform life-cycle rituals and other ceremonies for its members. Thus, the Sabha resolved to encourage *vaidika,* that is, religious education, and by 1916, it had disbursed a few prizes to students of good standing (Shri Subrahmanya Sabha 2008: 9). The necessity of English education, and indeed perhaps Western secular education, was also recognized. Scholarships were instituted, and despite periods of lull due to the indifference of the community, its support for poor students continued. From 1944, Kudpi Bhujanga Rao, a successful industrialist in Bombay, instituted a permanent trust of Rs. 20,000 to endow the Sabha with an annual sum of Rs. 600 for disbursing its scholarships. He went on to increase this sum gradually and also sponsored the construction of a student hostel in Mangalore (13). In parallel with the establishment and growth of the modern and 'secular' Sabha, the community also turned towards more traditional modes of organization. The Jnanashakti Subrahmanya temple in Pavanje near Mangalore originated in the desire of Kodimane Vasu Bhattaru to establish a temple run by the Sthanikas where they would not be discriminated against (Yaji Diwakar Bhat, pers. comm., April 2023).[20] It began construction in 1916 and was consecrated in 1920 (Sharma, Upadhyaya, and Udyavara 1982: 8). The complicated rituals that needed to be performed required the guidance of ritual specialists known as *tantris*. As the ritual status of the Sthanikas should make clear, they could not aspire to higher religious training in their own right and would have needed to employ Shivalli Brahmins to perform the rituals. However, there seem to have been a small number of *vaidika* families that produced priests. Kuraya Narayana Acharya had received training in the *agama shastras* outside the district, which allowed him to lead the consecration of the deity and institute daily worship to it.[21] The temple was encouraged by the Sabha, which began to hold its annual meetings on its premises (Shri Subrahmanya Sabha 2008: 10–11). A *pathashala*, or school for imparting traditional education in Sanskrit and priestly duties, was also opened. Sthanikas began to congregate in the temple to perform important life-cycle rituals such as the sacred thread ceremony and weddings. Today, the temple of Pavanje is the focal point for a small community of *vaidika* Sthanika families, who fulfil all the roles at the temple, from the *tantri* down to the ordinary priests (Yaji Diwakar Bhat, pers. comm., April 2023).

The *matha*, or monastery, and its *guru*, as the spiritual head of the community, are commonly found in Karnataka. The greater the antiquity of the *matha*, the greater is its prestige. Accordingly, the Sringeri *guru* is considered the highest seat of authority for the Smartas in South India, as the institution is said to have been established by Adi Shankaracharya, who is thought to have lived in the late eighth century CE (Lorenzen 1983: 156). At the beginning of the twentieth century, the Sthanikas found themselves without a *matha* of their own, which was one of the markers of Brahmin status. In addition, the *matha* had the powers to excommunicate, and conversely rehabilitate members of the sect. Efforts started in 1914 through appeals by letter to the *matha*, and in 1921, some leaders of the caste submitted a petition to the *guru*.[22] A delegation was sent to inquire into the conditions of the Sthanikas and their ritual observances. After it submitted its report, the Agent of the *matha* summoned the caste leaders. He informed them that they would be given a decree admitting that the Sthanikas were indeed 'Dravida Brahmins', were *shatkarmis*, that is, followed the six rites enjoined for Brahmins, and that there was no bar on interdining with them.[23] The *guru* handed over the decree on Makara Sankranthi, 1924. The immediate implication of the declaration was, of course, to finally grant the status of Brahmin, though this was scarcely respected in their own district. However, the bigger import of this campaign was to mark the Sthanikas as part of a larger religious community, which carried great prestige. Even if the Sthanikas had claimed to be Smarta Brahmins, now there were definite links that they could establish with other Smartas and with the teachers and followers of *Advaita* philosophy. The importance of the *guru* can be seen in the close contacts that the caste and its associations try to maintain with the *matha*. Thus, they achieved legibility in terms of a sectarian identity that is recognizable across Karnataka and South India.

Conclusion

It is likely that oral narratives of conflict between Smartas and Madhvas or Shaivas and Vaishnavas existed around Udupi. M.S. Achyuta Sharma might have drawn from this tradition and combined it with general claims by Sthanikas on temples of the region to position the Sthanikas as victims of a past injustice. However, this articulation was made possible by the sustained efforts of the Sthanikas in organizing themselves and their success in being recognized as Smartas. Though Sharma's 'history' of Udupi is patently a caste polemic, it deploys the concept of caste in a subtle manner. Instead, there is a packaging of caste identity with sect and religion to create an alternative but potent social imaginary. The Sthanikas are described as both Shaivas and Smartas, in their repeated description as 'Shiva-brahmins' and as disciples of Sringeri, respectively. This dual identity allows Sharma to speak as a representative of the Shaivas when he alleges that Shiva had stopped receiving worship in the temples, but also allows him to take the guise of a follower of large-hearted *Advaita* reproaching petty-minded Madhvas. The historical argument is made that the Sthanikas were the legitimate owners of the site of Udupi. Their fortunes had plummeted due to their opposition to the ascendant Vaishnavas, and later, to the British. Despite the defence of Shiva, Sharma does not seem to disparage Krishna or Vishnu. Thus, it is not a straightforward sectarian polemic, though it makes references to an unstated and presumed history of violent clashes between Shaivas and Vaishnavas. Sharma instead pitches his protest as an insistence that orthodox precepts related to worship be followed. He concedes that forms for a formless and eternal God are imagined (Sharma 1969: 75).[24] Yet, in this material world, specific deities need to be worshipped and only by those who believe in them.[25] 'Even if the government is the same, would it be proper for the executive and the judiciary to meddle in each other's duties?', he asks (75). He then calls on the government of the day to protect temples by nominating as trustees only those who were believers in its deities (75). In this articulation, he appears to speak for all Hindus, having presented *Advaita* as its true essence and himself as its votary. Presumably, this amplifies his critique of the antagonists of the Sthanikas by presenting their interests as inimical to that of all of Hinduism.

This social history presents an example of caste mobility in a decidedly upper-caste group. As a micro-study, its aim has been to present the trajectory of the Sthanikas through a detailed study of archives and by presenting them in the specific regional context. There is a clear contrast between colonial representations of the Sthanikas in the nineteenth century and their self-representations in the twentieth century. In the former, they appear in the record as objects of an Orientalist ethnography, where the narrative of their origin seems to have piqued curiosity but was accepted at face value and never invited serious study. The Sthanikas, on the other hand, were not simply an artefact of the colonial knowledge apparatus but had a concrete existence. The idea that a natural community existed, either as ethnography would have it in terms of a textual delineation or based on kinship ties, is simplistic. The creation of an identity and a community based on it demands criteria for inclusion and exclusion. Here, narratives have played an important role in Sthanika discourses on caste. History is central to these narratives, and is invoked repeatedly, since the past was thought to be quite different from the present, and its evolution into the present needed to be explained. Their admission into a larger religious community of the Smartas and its related self-fashioning also enabled a different type of critique with its own attractive claim to historicity. Through the example of a contested history of the temple town of Udupi, we have seen the appropriation or at least borrowing of a powerful polemic involving violent sectarian conflict. In the twentieth century, the Sthanikas thus emerge not only as subjects of caste and religious communities but also as subjects of history.

Notes

1 The tradition of Parashurama is found in coastal Karnataka and Kerala, appearing in texts such as the *Sahyadri Khanda* in Sanskrit, *Gramapaddhati* in Kannada and *Keralolpatti* in Malayalam.

2 These are Udupi, Subrahmanya, Kumbhasi, Koteshvara, Shankaranarayana, Gokarna and Kollur.

3 Inscriptions from nearby Udyavara contain imprecatory verses that condemn the destroyer of the inscription to the sin of destroying Varanasi and Shivalli.

4 In this way, what might have been an oral narrative about the origins of the Sthanikas was related to Sanskrit texts, namely the Manusmriti, as a sanctioned alliance of a lower stature.

5 The Krishna *matha* is so called because it is not strictly a temple according to the *agama shastras*. Nevertheless, it is a shrine and thus a temple in the general sense of the term.

6 These are the Sode, Adamar, Phalimar, Pejavara, Krishnapura, Kaniyur, Puttige and Shiruru *mathas*. These villages are in the coastal tract except for Sode which is above the Western Ghats.

7 Despite the critique of colonial ethnographies, they are indispensable as sources for such information. Greater confidence may be placed in descriptions of a prominent group such as the Brahmins.

8 The Kota Brahmins, one of the Tulu Brahmin castes are Smartas but do not belong to any *matha*. Instead, the deity Narasimha in the temple at Saligrama in Udupi district is considered the *guru*.

9 In the novel *Samskara* by the Kannada novelist UR Ananthamurthy which is set in an *agrahara* or an orthodox Brahmin village, Smarta characters are amused by the ritual purity of their Madhva neighbours and gently mock their orthodoxy.

10 For instance, a draft was drawn up of a proposed amendment to the district manual for its section on the Sthanikas (Shri Subrahmanya Sabha 1943: 1–5).

11 An Ayurveda doctor, he is alleged to have been an underground Congress worker in the 1940s. He is remembered as the only caste leader who agitated for the Sthanikas and confronted the powers that be. These include activities as a trustee of the Kukke Subrahmanya Temple to which he was selected in 1953, petitions to courts and a few controversial writings. See the biographical sketch 'Dr. M.S. Achyuta Sharma, L.I.M.' (Shri Subrahmanya Sabha 2008: 40).

12 The author claims that he was compelled to write this book since there was no proper response to a pamphlet he had distributed about certain errors in the conduct at Udupi during the *paryaya* celebrations of 1968. *Paryaya* refers to the festivities on the occasion of the priesthood of Krishna changing hands every two years.

13 While Ananteshwara may be interpreted as an epithet of Shiva meaning 'the infinite lord', Anantasana which means 'seated on Ananta' would refer to Vishnu, seated on the serpent Anantashesha.

14 A *saligrama* stone is a pebble containing fossilized ammonites. Due to the similarity of the spirals to those on Vishnu's *chakra* or discus, the stones are worshipped as representing Vishnu (Thurston 1909b: 323).

15 Sharma cleverly uses a pun here. *Abheda* stands for Shankara's monism and *bheda* for dualism, the principle of Madhvacharya. However, in colloquial speech, *bhedavada* means discrimination.

16 This is a reference to the rebellion of Kalyanaswamy that originated in Coorg, a short-lived armed insurrection in 1837 (Bhat 1998: 165–87).

A central role is claimed for one of the three *dewans* of Coorg, Lakshminarayana. See the biographical sketch 'Dewan Lakshminarayanayyavaru' (Shri Subrahmanya Sabha 2008: 35).

17 The Shri Subrahmanya Sabha celebrated its centenary in 2008 when it released a commemorative volume containing invaluable information about the organization as well as the community.

18 The records of the early years are lost, but these details had been assembled by the organizers of the golden jubilee in 1959 by reaching out to elders within the caste and by requesting them to hand over any scraps of paper that might be relevant (Shri Subrahmanya Sabha 2008: 8).

19 A short biography of Krishnayya provides a few interesting details. He was born in 1859 to Kumbhat Sankayya and Lakshmi in the village of Kadaba near Kukke Subrahmanya, the principal seat of serpent-worship in the district, on which the Sthanikas lay a claim. The Kumbhat family claims to have had a history of at least 500 years as 'prime ministers' and *shanbhogues* to the local Kadamba chiefs. Sankayya was the *magane shanbhogue* (accountant for a village cluster) and the family seems to have controlled four temples around Kadaba, where Sthanikas themselves served as priests. Krishnayya trained to become a *shanbhogue* himself (Rao 2012: 6).

20 As with Kumbhat Krishnayya, tradition attributes a long history to the family, which claims to have established the ancient temple of Mahalingeshwara on the hillock in Pavanje. In the recent past, male members of the family had been serving the temple as *parupatyagara* or supervisor. Ramappayya was serving in this capacity when during the annual festival, he felt insulted by the Shivalli Brahmin priests of the temple. After all the other temple servants had been given the *prasada*, the priest supposedly called for the 'sanika' to appear and threw it into his cupped palms from far away. Smarting from the humiliation, he vowed to never step foot inside the temple, and started collecting funds for the construction of a temple that would be exclusively managed by the Sthanikas. This personal mission would be completed by his son Vasu Bhattaru (Yaji Diwakar Bhat, pers. comm., January 2023).

21 See the biographical sketch 'Ve. [Vedamurti] Sri Kuraya Narayanacharyaru' (Shri Subrahmanya Sabha 2008: 37–8).
22 See 'Adhyaksha Bhashana' (Shri Subrahmanya Sabha 2008: 17–19).
23 Sometimes, the Brahmins conceded that the Sthanikas were *trikarmi* Brahmins, that is, followers of three rites but not all the six rites of the Brahmins and thus were a lower class of Brahmins. This accords well with Buchanan's description. (Yaji Diwakar Bhat, pers. comm., January 2023).
24 It is clear from other passages in the text that God is equated to *Brahman* of *Advaita*.
25 Shankara is said to have propounded that the only path to salvation was wisdom, which was essentially the realization of the unity of *Brahman* and *Atman*. However, 'for the ordinary man in the world devotion is a valuable propaedeutic to raise himself to a higher level but cannot in itself produce liberation' (Lorenzen 1983: 159, 172).

References

Adiga, Malini. 2001. 'Shivalli Brahmanas: A Historical Analysis of Their Origins'. In *Shivalli Brahmanaru: Collection of Articles on Shivalli Brahmins Containing Historical, Social and Literary Aspects*, edited by HBL Rao, 56–81. Mumbai: Shivalli Pratishthan.

Arnold, David, Robin Jeffrey and James Manor. 1976. 'Caste Associations in South India: A Comparative Analysis'. *The Indian Economic and Social History Review* 13(3): 353–73.

Bayly, Susan. 1999. *Caste, Society and Politics in India from the Eighteenth Century to the Modern Age*. Cambridge: Cambridge University Press.

Bhat, N. Shyam. 1998. *South Kanara (1799–1860): A Study in Colonial Administration and Regional Response*. New Delhi: Mittal Publications.

Buchanan, Francis. 1807. *A Journey from Madras through the Countries of Mysore, Canara and Malabar Vol. III*. London: T. Cadell & W. Davies / Black, Parry & Kingsbury.

Chatterjee, Partha. 2008. 'Introduction: History in the Vernacular'. In *History in the Vernacular*, edited by Raziuddin Aquil and Partha Chatterjee, 1–24. Ranikhet: Permanent Black.

Cohn, Bernard S. 1996. *Colonialism and Its Forms of Knowledge: The British in India*. Princeton, NJ: Princeton University Press.

Dirks, Nicholas B. 1996. 'Foreword'. In *Colonialism and Its Forms of Knowledge: The British in India*, ix–xvii. Princeton, NJ: Princeton University Press.

Jaffrelot, Christophe. 2000. 'Sanskritization vs. Ethnicization in India: Changing Identities and Caste Politics before Mandal'. *Asian Survey* 40(5): 756–66.

Lorenzen, David N. 1983. 'The Life of Sankaracarya'. In *Experiencing Siva: Encounters with a Hindu Deity*, edited by Fred W. Clothey and J. Bruce Long, 155–75. New Delhi: Manohar Publications.

Nanjundayya, H. V. and L. K. Ananthakrishna Iyer. 1928. *The Mysore Tribes and Castes Vol. II*. Mysore: The Mysore University.

Rao, K. Gopal. 2012. *Kadaba Kumbhat Krishnayya*. Kadaba: Navarasa Sahitya Male.

Rao, Nagendra. 2005. *Brahmanas of South India*. Delhi: Kalpaz Publications.

Rao, Vasudeva. 2002. *Living Traditions in Contemporary Contexts: The Madhva Matha of Udupi*. New Delhi: Orient Blackswan.

Rudolph, Lloyd I. and Susanne Hoeber Rudolph. 1960. 'The Political Role of India's Caste Associations'. *Pacific Affairs* 33(1): 5–22.

Sanyal, Hitesranjan. 2022. 'Social Mobility in Bengal: Its Sources and Constraints'. In *Caste in Bengal*, edited by Sekhar Bandyopadhyay and Tanika Sarkar, 40–92. Ranikhet: Permanent Black. (Reprinted from Hitesranjan Sanyal. 1981. *Social Mobility in Bengal*, 33–64, 113–23. Calcutta: Papyrus).

Sharma, M.S. Achyuta. 1969. *Sri Udupi Kshetrada Naija Chitra Mattu Charitrika Hinnele (The True Historical Picture of Udupi)*. Mangaluru: Satyashodha Prakashana.

Sharma, P. S. V., N. V. Upadhyaya and V. R. Udyavara. 1982. *Pavanje Sri Jnanashakti Subrahmanyakshetra*. Mangalore: Belle Bhavani Rao.

Shri Subrahmanya Sabha. 1942. *Memorandum of Association*. Mangalore: Shri Subrahmanya Sabha.

Shri Subrahmanya Sabha. 1943. *Draft of Proposed Amendment: Sthanikas*. Mangalore: Shri Subrahmanya Sabha.

Shri Subrahmanya Sabha. 2008. *Nooru Hejje: Shatamanotsava Smarana Sanchike (A Hundred Steps: Centenary Commemorative Volume)*. Mangalore: Shri Subrahmanya Sabha.

Srinivas, Mysore Narasimhachar. 1972. 'Sanskritization'. In *Social Change in Modern India*, edited by Mysore Narasimhachar Srinivas, 1–48. Hyderabad: Orient BlackSwan.

Stoker, Valerie. 2016. *Polemics and Patronage in the City of Victory: Vyāsatīrtha, Hindu Sectarianism, and the Sixteenth-Century Vijayanagara Court*. Oakland, CA: University of California Press.

Sturrock, John. 1894. *Madras District Manuals: South Canara, Volume I*. Madras: Government Press.

Thurston, Edgar. 1909a. *Castes and Tribes of Southern India, Vol. I*. Madras: Government Press.

Thurston, Edgar. 1909b. *Castes and Tribes of Southern India, Vol. II*. Madras: Government Press.

Thurston, Edgar. 1909c. *Castes and Tribes of Southern India, Vol. VI*. Madras: Government Press.

3

Culture, Identity, Religiosity: The Ideal Tuluva in Tulu Cinema

Vishvajith Madhavamurthy and Anubhav Sengupta

Introduction

The Tulu sub-nationalist movement, once considered dead, has resurrected in the past two decades. It aims to fetch Tulu – a language natively spoken by about two million speakers[1] – the status of a scheduled language in the 8th Schedule of the Indian Constitution and as an official language of administration in Karnataka. It also aspires to build a Tulu-speaking Tulunad state comprising the Dakshina Kannada and Udupi districts of Karnataka and the Kasaragod district of Kerala. The demand for a Tulu-speaking linguistic state was first articulated in the 1950s by Tulu intellectuals like S.U. Paniyadi, in opposition to the region's inclusion with either Kerala or Karnataka, as demanded by the sub-nationalist movements in those regions (Paniyadi 2007: 139). The Tulu movement, however, failed to gather enough political support and remained dormant until now.

The Tuluva identity is increasingly asserting itself, driven by fears of a dwindling Tulu-speaking population, a general perception that the Tulu language has been reduced to a 'secondary status' even within Tulunad, and a lack of awareness of the language and its script among Tuluvas, which have led to efforts to revive the script (Gururaj and Kulal et al. 2022: 186). While the resurgence of the movement was initially viewed as a 'language renaissance' (Someshwar and Hiriadka et al. 2000: 85), the wide use of political symbolisms, including the

Tulu flag, the deified Tulu mother-goddess Tulu *Appe*[2] and the rise and involvement of multiple cultural and political organizations like Tuluvera Paksha,[3] Tulu Rakshana Vedike,[4] TuluWorld Foundation[5] and Jai Tulunad, among others, demonstrates the political nature of the movement.

The term 'Tuluva' (trans. Tulu people) has been used interchangeably to refer to the language, the people and the region (Caldwell 1875: 6–8), indicating that it is a complex identity and not merely linguistic. Allegations of communalization of the Tuluva identity and the Tulu movement persisted even in the 1950s and had to be defended by asserting its linguistic nature (Paniyadi 2007: 139). Despite claims of cultural syncretism in Tulunad, the Tulu movement might eventually encounter questions of inclusion and exclusion of identities, communities and peoples, even as it foregrounds the Tuluva identity. This chapter examines the nature of the culturally produced Tuluva identity on the grounds of inclusion and exclusion and seeks to establish that this culturally produced identity is a complex identity built on particular and exclusive notions of caste, gender, faith and associated traditions. The present assertions of the Tuluva identity and the demand for statehood by the Tulu movement in a multi-lingual, multi-cultural region become significant given the inherent characteristic of nationalism to homogenize identities.

Nationalism and the Ideal Subject in Cinema

Nationalism requires subjects to abandon their particular identities and adopt a shared, homogenized identity (Grotenuis 2016: 109–10). While this homogenization of identities occurred due to the promise of equality in the West, postcolonial nations produced exclusionary national ideals built on the 'gold standard' of dominant identities and ideologies (Grotenuis 2016). The ideal subject is considered a legitimate representation of the national ideal. It is simultaneously a product of nationalism that embodies nationalist ideals and a tool that reproduces nationalism itself (Zake 2002: 218), signifying its centrality to the project of nationalism. The national ideal often prevents other communities and identities from equal participation in nation-building and citizenship, creating the citizen-subject dichotomy and subjectivizing those excluded by the ideal.

The national ideal is a culturally produced identity for its subjects to emulate; nationalism utilizes 'cultural products' and transforms them into 'substantial representations of national essence' for the national(ist) subjects to interpret as 'manifestations of the uniqueness of the ethnic group or nation' (Zake 2002: 333). By doing so, nationalism provides the national subjects with an environment of 'national(ized) experience' in which the subject can 'authentically' exist (Zake 2002). Mass media is one of the ideological state apparatuses contributing to the 'socialization of individuals into becoming subjects' (Zake 2002, 222). One of the key mediums that constitute mass media is cinema.

Cinema, particularly talkies cinema, does not require literacy to access it, unlike print, which makes it an effective medium for imagining a nation vis-à-vis print (Raghavendra 2021: 91–2). National cinema, therefore, was a 'key project' of postcolonial nations, particularly in its popular or commercial form in countries with low literacy levels like India (Raghavendra 2021). Combining language with visual and aural symbols, cinema provides a 'standardized, shared system of reference', aiding in people's self-identification as individuals, communities and nations (Sorlin 1996: 10).

In India, popular Hindi cinema was constructed and addressed 'wide audiences within the boundaries of' the Indian nation, and it 'deliberately tailored' itself to the role of national cinema (Raghavendra 2011: xi). It 'served political purposes' of producing a 'homogenous "Indian" culture' and defining 'the nation' (Raghavendra 2011, xxxv). In addition to being a tool to imagine the nation, national cinema foregrounds an abstract ideal subject through the nationalist ideal to address the nation as its constituency. The abstract Indian ideal subject produced by Hindi cinema, which reflected the dominant secular ideology, was deconstructed as an upper-caste

Hindu-male 'normative' subject (Bhrugubanda, 2018: 5). Popular Hindi cinema has also been studied sociologically to deconstruct the specific ideals that represent the 'Indianness' of Indians projected in films, including ideas of 'ideal' Indian women (Uberoi 1998: 305–33). However, popular Hindi cinema was not the only popular cinema in India.

India saw another level of political organization – the sub-national, which resulted in the emergence of 'regional' cinema alongside the 'national', configuring the sub-national and reconfiguring the markets based on the 'regional languages'. Regional language cinema addressed their linguistic constituencies rather than a universal subject (Raghavendra 2011: 99). In other words, these cinemas, mainly from South India, reconfigured the national subjects produced by Hindi cinema as a dual national–sub-national subject by projecting their version of nationalist ideals. For instance, Kannada cinema produced its own ideal subjects and projected the national ideal for the Kannada nation. For example, against the backdrop of the threat to the Kannada nation from the Kodava community, who refused to be identified as Kannadigas, Kannada cinema projected one specific type of Kodava as the ideal for the Kodava subject to emulate – the ideal Kodava is a male heroic warrior, true to his traditional role, and is available at once at the service of the Indian nation and the Kannada nation. The ideal Kodava is the one who embraces and upholds the Kannadiga identity and cannot merely be a Kodava (Dechamma, 2010: 202–4). In line with such studies of identities and ideals culturally produced in films, this chapter inquires into the characteristics that define the 'Tuluvaness' of the Tuluvas by analysing and deconstructing the ideal Tuluva identity.

Sociological attention to the category of 'national cinema' has traditionally treated it as a mass-media social communication tool and examined how the medium generates 'messages', the ability to understand them, and determines who gets included or excluded in the 'moral and economic community' called the nation (Gellner 1983: 127). However, rather than the medium itself, it is the 'content' that is of 'cardinal interest' for the cultural industries that produce cultural products like cinema, which is precisely why policies are more concerned about the content – that is, what is portrayed and how – rather than the mechanisms of transmission of such content (Schlesinger 2000: 20). However, being a work of fiction, sociological research has avoided the analysis of narrative cinema and has only paid attention to it in the past few decades (Uberoi 1998: 306). Departing from the tradition of film studies, a sociological approach to reading films focuses on the film narratives over the formal aspects of cinema to extract the embedded ideology in popular films. Popular cinema, it has been argued, constructs an 'ideal moral universe' that is 'intrinsically' connected with 'ideas of tradition and nation' (Uberoi 1998). Cinematic narratives of popular films can be seen as a source of data that can 'open up dimensions of human experience' despite the 'conceptual and methodological pitfalls involved in such exercises' (Uberoi 1998).

This chapter investigates the Tuluva identity based on a narrative analysis of the cinematic narratives of select popular Tulu films, that is, films that are focused on Tulu people and culture. The films analysed in this chapter include *Kanchilda Baale* (2011), *Barsa* (2016), *Yesa* (2017) and *Kantara* (2022). Despite being a Kannada-language film, *Kantara* can be considered a Tulu film for all practical purposes, given that its narrative is focused on Tulu people and culture. Production-wise, it involved many cast and crew members from Tulunad and incorporated Tulu songs and dialogues. Its immense popularity in Tulunad and the subsequent demands by a section of Tuluvas to dub the film in Tulu – a rarity in itself,[6] claiming that the Tulu-dubbed version would be 'an ideal representation of' the land and that 'the nativity of the film and Tulu' would go hand in hand[7] with the *daivas*[8] speaking the divine language,[9] that is, Tulu, justifies its consideration as a Tulu film.

The chapter will first establish that the Tuluvas were denied their identity in Kannada films – films produced by the Kannada nation that ideally should have included its minorities, including the Kodavas and the Tuluvas. The chapter will then map the recent developments in Kannada cinema, which demonstrate a renewed interest in the Tuluva identity. It will then present a narrative analysis of the films listed earlier and establish that the ideal Tuluva, being culturally produced in cinema, is a devout figure who upholds his faith at all costs and is an exclusionary figure based on the sociological categories of caste, class, gender, language and religion.

Invisibilized Tuluva Identity in Kannada Cinema

Paniyadi's assertion for a Tulu state was based on the fears of such invisibilization arising from the self-identification of Tuluvas as Kannadigas outside the Kannada-speaking regions and the invisibilization of Tulu and the Tuluva, given the influence of Kannada culture backed by the Basel Mission,[10] alongside the adoption of Konkani, Hindustani or Malayalam as mother tongues by Tuluva converts to Christianity and Islam (Paniyadi 2007: 139–40). Such an invisibilization of the Tuluva identity has been noted in Kannada cinema as well.

The Tuluva was traditionally portrayed as a Kannadiga in Kannada cinema, invisibilizing their 'Tuluvaness'. Kannada films subjectivized non-Mysore dialects, languages and cultures through subordinated representations and characterizations by assigning non-Old Mysoreans to roles that provided comic relief and by privileging and prioritizing Old-Mysore dialect and culture (Raghavendra 2011: 13).[11] Such invisibilization of Tuluvaness and subordination of the Tuluva identity, among other regional identities like the Kodava identity, persisted until recently, as seen in films like *Pancharangi* (2010). Typically, such films reduce cultural differences in dialect and appearance,[12] only to expose the Tuluvaness of the characters in comical situations when the Tuluva momentarily switches from Mangaluru Kannada to Tulu.[13]

In the last decade, however, several Kannada-language narrative films like *Ulidavaru Kandanthe* (2014), *Ondu Motteya Kathe* (2017),

Garuda Gamana Vrushabha Vahana (2021) and *Kantara* (2022) have gained popularity in Karnataka, including the Tulunad region, for their Tuluva-Tulunad-centric narratives and representations. Made by the three Shettys of Kannada cinema[14] from Tulunad, these films have inverted the Old-Mysore-centric narratives of Kannada cinema, foregrounding the linguistic and cultural diversity of the Tulunad region through realistic representations. These films, typically made in the Mangalorean/Udupi dialect of Kannada,[15] with a liberal mix of Tulu and Kundagannada,[16] played an important role in popularizing cultural markers of Tulunad, including *Pilinalike*, *Kambala* and *Daivaradhane* outside the region.

However, *Ulidavaru Kandanthe* and *Garuda Gamana Vrushabha Vahana* narrate seemingly universal, secular stories and use Tulunad and its culture only as a narrative backdrop, in addition to their embedded aesthetic values. Overt assertions and pride concerning the Tuluva identity are absent in such films, given their secular themes of crime, romance, friendship, etc. For example, Richie, one of the critical characters in *Ulidavaru Kandanthe*, hails from a Christian family, invokes cultural references to *Jumaadi*[17] and *Banta*[18] – deities or *daivas* from the *Daivaradhane*[19] belief system – and calls the land he inhabits the land of *dharma*[20] and *punya*.[21] Neither does Richie overtly invoke the identities of Tuluva or Tulunad, nor does the reference to the *daivas* play any significant role in the narrative. This changed with *Kantara*.

Kantara placed *Daivaradhane* at the centre of its narrative and included multiple cultural markers associated with Tulunad. The Tuluvas were divided over the portrayal of the *daivas*, including in future films like *Kantara 1* (2025) and *Kaljiga* (2024), as *Kantara*'s fame produced outcomes such as mockery, emulation[22] and even spread of faith outside Tulunad[23] – all violations of the basic tenets of the faith. *Kantara*, however, is not the first film to represent *daivas* on the big screen. Tulu films like *Koti Chennayya* (1973), *Koti Chennayya* (2007), *Gaggara* (2008), *Kanchilda Baale* (2011) and *Paddayi* (2017), among others, have depicted the *daivas* and *Daivaradhane* in the past to the extent that such repetitious depiction of Tulunad's cultural markers has received flak for catering to the oriental gaze of the Kannadigas[24] and for driving away Tulu audiences towards 'mindless' comedies (Mangaluru 2007: 997). Contrarily, a study on Tulu film viewership indicates that while they enjoy Tulu comedies, a substantial section of Tuluvas appreciates the representation of its culture the way *Gaggara* and *Kanchilda Baale* do (Darshan & Prasad 2020: 9263). These opposing viewpoints are representative of the Tuluvas' ambivalence towards *Kantara* and such non-Tulu films that depict *Daivaradhane* in both visual and filmic narratives.

Despite these developments, barring the reading of *Kanchilda Baale* (2011) and *Gaggara* (2008) as ethnographic sources,[25] there is a dearth of academic research that engages critically with narratives of films concerned with Tulu culture and identity. Tulu films are gaining popularity[26] and are increasingly becoming sites of political symbolism and assertions of the Tuluva identity.[27] The centrality of cinema and the role and impact of representation of the Tuluva culture in both Tulu and Kannada films to the current social and political identity-based conflicts provide a solid point of entry for this chapter to inquire into the nature of representations and self-perception of the Tuluva identity in such films.

The Culturally Produced Ideal Tuluva

Tulu films, and even films made in Kannada but with a strong sense of Tuluva identity, like *Kantara,* produce and project an ideal Tuluva identity. This ideal represents a set of values a Tuluva must have, irrespective of caste, gender, faith and language. The ideal is built on moral values that uphold religiosity, respect faith and preserve tradition. This ideal, as this section shall demonstrate, reflects the dominant caste's view of how society must be structured to protect its interests.

The Ideal Tuluva and Religiosity

The Tuluva identity is built on traditional practices rooted in the feudal social structure. *Daivaradhane* is the foundational element that holds this structure together. *Daivaradhane* is inviolable and sacrosanct. Violation of faith is unacceptable, and every effort must be made to preserve it. The ideal Tuluva demonstrates religiosity in daily life, irrespective of any other complementary faith they may follow. They find recourse to faith if they have drifted away from faith and tradition.

Kantara narrates the story of a village located amidst a sacred forest that was once handed over by a feudal king to the villagers in exchange for eternal peace for his clan, on the condition that the villagers be allowed to dedicate themselves to the service of the deities – *Panjurli*[28] and *Guliga*.[29] The conflict within the narrative is between the villagers and the descendants of the king – a landlord from Mumbai referred to as *Dani* (trans. 'landlord') and later his son, Devendra Suttur – both of whom attempt to recover the land so that they can monetize it for their interests. The film's protagonist, Shiva, belongs to a family of *daivanarthakas* – *Bhuta* impersonators drawn from select Dalit sub-castes who impersonate the *Bhutas/Daivas* and are 'possessed' by them during the *Bhuta Kola*[30] ceremony. He refuses to take on the hereditary role of *daivanarthaka* after watching his father disappear into the forests during a *Bhuta Kola* ceremony. Shiva consistently violates his faith by hunting wild boars[31] and cutting forest trees illegally for timber to benefit his feudal master. The story deals with Shiva's journey of self-realization and uprising as a protector of forests and as an upholder of tradition. Towards the end, Shiva not only fights his feudal master head-on but also ends up taking on the responsibility of performing as a *daivanarthaka*.

In *Kantara*, the key antagonist, Devendra Suttur, and his father, Dani[32] – who return from Mumbai at the film's beginning to demand the return of land from *Panjurli* – violate tradition and faith. By establishing their rights patrilineally, they disregard the traditional matrilineal inheritance system of *Aliya Santana*, which 'passes landed property to the maternal nephew instead of one's progeny' (Benjamin 2017: 205). By choosing the court over the word of the *daiva* – the spirits of *Daivaradhane* – Dani not only violates the ancestral promise made to the *daiva* but also questions the *daiva*'s judicial jurisdiction. Differentiating the *daivanarthaka* from the *daiva* by both characters is not a mere insult to a particular *daiva* or the *daivanarthaka*; it is akin to questioning the very basis of *Daivaradhane*. Dani and Suttur represent Tuluva non-believers; the punishment meted out to the nephew by *Panjurli* and to Suttur by *Guliga* points to the unpardonable nature of their crime, the inviolable nature of faith, and the sacrosanct status of *Daivaradhane* within Tulu culture.

Crucially, even the forest officer in *Kantara*, Muralidhar, is a non-believer. As an individual, Muralidhar is a non-Tuluva outside the framework of faith. One may, in this case, extend film theorist M.K. Raghavendra's methodological intervention to interpret archetypal film characters like police, judiciary, etc., as allegories of the state (Raghavendra 2021: 94) to the forest department. As a forest officer, Muralidhar represents the state. *Kantara* appeals to the individual outside faith and the state to respect the Tuluvas' faith system, voluntarily participate, and partner with them through mutual understanding to uphold faith and tradition and treat nature as sacred. Faith and secular thought may have different approaches, but their aim is one. Yet, it is beneficial if the secular forces empathize with faith rather than imposing secularity.

The ideal Tuluva projected by *Kantara*, therefore, is a Tuluva who places faith in *daivas*, recognizes the sacrosanct nature of *Daivaradhane*, and ensures that they neither question the jurisprudence and judicial authority of the *daivas* nor violate faith. The Tuluva, who steps out of faith for various reasons but does not violate it, can always become the ideal by opting to revert to faith. This religiosity of the Tuluva, the unquestionable faith in the *daiva*, the inviolability and sacrosanct nature of *Daivaradhane*, and the recourse for the non-believer Tuluva are recurring themes across multiple Tulu films.

Kanchilda Baale (2011) narrates how a Tuluva family deals with questions of survival of faith, continuity of tradition, superstition and rationality. Jaanu and her parents return from the United States on vacation to her great-grandmother Ajjamma's house in Tulunad. Manu – Jaanu's father and Ajjamma's grandson – is a doctor who intends to build a hospital in the village to benefit its poor and oppressed. The film's narrative revolves around the ideological clash between a rationalist Manu – who rejects archaic traditions, and Ajjamma – who wishes to fulfil her Kanchil vow to the *daivas*. The practice of *Kanchil* depicted in *Kanchilda Baale* is a celebration of the girl child, who is crucial for *Aliya Santana*. Not having a girl child ends the family lineage and also violates Aliya Santana's matrilineal inheritance (of traditions and property) system. Rationality cannot come in the way of faith and tradition.

Ajjamma asks her son, Manu's father Mahabala, to arrange a *Kola* so that she can fulfil the *Kanchil* vow. She reveals to Manu and his wife that when the couple got married but did not bear a girl child for four years, she promised the *daivas* to perform the *Kanchil* ritual upon being blessed with one who would carry forward her lineage, which offends Manu and his wife. An argument ensues between the couple and Ajjamma:

> Manu: 'When we got married, my wife was still studying. We decided to delay having kids by 4–5 years until she finished her studies. Given that, there's no meaning in claiming that the girl child was born because of the vow.'
>
> Ajjamma: 'You will say that it was your wish that you delayed having a child, but how is it that a girl child was born? Isn't that the wish of our daivas?'
>
> Manu: 'I don't have any issues with conducting the Kanchil ritual to celebrate, but my conscience cannot accept that the girl child was born solely because of the vow.'

Despite efforts made by his wife, daughter and a dejected Ajjamma to convince him to conduct the ritual, Manu remains firm in his rational stance. Later, in a narrative sub-plot, a *Maadira*[33] singer arrives at their doorstep and sings the *paddana* of Maayinda Mani at Ajjamma's request. The *paddana* narrates the story of the defiant toddy-tapper Aali Baali, who refuses to contribute offerings to a *Kola* for *Jumaadi daiva* – drawing parallels between Manu's actions and those of Aali Baali. *Jumaadi* punishes Aali Baali by making his niece and her son disappear. At the end of this sub-plot, Ajjamma tells the entire family:

> What we learn from this story is the manifestation of the Bhuta in a family, what happens when a promise made to the Bhuta is fulfilled, and, thirdly, that the Bhuta will never leave those who have faith in them.

Manu does not respond, and his stance remains unchanged. The rest of the story deals with the consequences of this refusal. These consequences come in the form of Jaanu's disappearance on the night of the *Kola,* when Kaalu – the house help – attempts to assault her sexually. The timely intervention of *Jumaadi Daiva* saves Jaanu. This divine intervention is attributed to the faith that the clan had for generations, as Ajjamma says:

> The daiva saved my girl. I told everyone, 'Let's fulfil the Kanchil vow for the girl.' Did anyone listen? Now, look at her plight.

Ajjamma instructs Mahabala to arrange a *Kola* within four days to fulfil her vow, adding that it is her 'final word' to which 'nobody can object'.

Manu responds:

> Yes, Ajjamma. I had decided to fulfil the *Kanchil* vow when I found the girl safe.

Manu willingly conducts the *Kanchil* ritual for *Jumaadi.* After the ritual, *Jumaadi* arrives at their doorstep and assures the family that they will always be protected by him, given their faith in him.

Manu's recourse to faith is similar to *Kantara*'s Shiva's spiritual journey. Manu and Shiva are not disrespectful towards the *daiva*, unlike Aali Baali, Dani, or Suttur, and therefore can adopt the ideal. Manu's return to faith from rationality, however, does not mean a rejection of modernity. Manu fulfils his dream of building a hospital to rescue the oppressed illiterates at the mercy of superstition. Faith and rationality have their own demarcated spaces. Rationality is best used for the benefit of society, not to question faith.

Tulunad is home to multiple faith systems, including Islam, Christianity and Brahminical Hinduism. It is common for a 'Hindu' Tuluva to follow a 'dual religious life', worshipping both the 'Puranic Gods of the Hindu pantheon' and the *daivas* (Upadhyaya and Upadhyaya 2002: v). These faith systems are complementary, not competing, in the daily lives of a Tuluva. Tulu films are reflective of this reality. However, there appears to be an inherent hierarchy that places *Daivaradhane* over and above other faith systems. The films often show mainstream Hindu gods through photo frames placed in households and as temple deities. Yet, *Daivaradhane* emerges as the 'eternal reality' and 'true faith' by categorizing mainstream practices like *Homa, Vaastu,* astrology, etc., as 'superstitions'. *Kanchilda Baale* introduces a Brahmin character named Kuppanna Bhat to serve this hierarchization and present a critique of 'superstition'.

Manu learns from Koraga – a low-caste farm labourer who works on his ancestral land – that the village astrologer, Kuppanna Bhat, has been financially exploiting him under the pretext of exorcising the spirit of his dead son, Choma, who has possessed Koraga's daughter. Manu confronts Bhat and alleges that astrologers continue to exploit 'illiterates' despite the 'scientific progress' made by the country. Alleging that Bhat has neither 'exorcised' Choma's spirit nor 'healed' Koraga's daughter, Manu tells Bhat that he will treat Koraga's daughter and takes away the cattle, which Bhat had forced Koraga to 'donate' as payment for the ritual expenses. To avenge this insult, Bhat attempts to extract money from Ajjamma on the pretext of conducting rituals that can change Manu's stance regarding the *Kanchil* ritual. Bhat does not appear in the narrative again. However, when Manu returns to faith towards the film's ending, an on-screen textual message to the audience reads, 'Save the original faith, discard superstitions',[34] not only to critique irrational rituals as 'superstitions' but also to uphold *Daivaradhane* as 'true faith'.

Notably, anxieties about Brahminical rituals taking over the rituals of the 'true faith' of *Daivaradhane* and wielding power over even Christians and Muslims, particularly in matters of land purchase, are real. Solomon Benjamin's study on the nature of urbanization in the region reflects these anxieties while establishing 'sacred territory' beliefs that apply to all faiths in the region (Benjamin 2017: 208). The ideals projected in *Kanchilda Baale* provide recourse to faith and assert *Daivaradhane* as the true faith. Most Tulu films, however, leverage *Daivaradhane* even when faith is not a core concern within the films' narrative.

Barsa (2016) is a run-of-the-mill action-comedy that stitches together multiple comical subplots and an action-romance storyline. It narrates how Prithvi, the lead male protagonist, helps the shopkeepers of a market save their shops from being demolished by the greedy builder Shivaram. Prithvi is a devout follower of *Koragajja*, a prominent *daiva*. His faith in *Koragajja* is so strong that the film leaves no room for coincidences, depicting all such events as a result of faith's intervention in his life.

Barsa opens with Prithvi's father, Vishwanath, performing the customary Friday morning *pooja* at the *daivasthana* within the house's premises. Yet, a young, school-going Prithvi is missing. Being a devotee of *Koragajja*, also called *Ajja*, he is seen worshipping *Ajja* at a shrine outside the house's premises. Prithvi's deep devotion is such that the priest who takes care of Ajja's shrine arrives at Prithvi's home under the pretext of conducting Ajja's *Kola* and tells Vishwanath:

> Vishwanath, your younger son Prithviraj has immense love and faith in the daiva. I've seen him; he comes to Ajja's shrine, keeps

> smiling, and talks to himself. You cannot find such dedication and faith among children these days, even if you look for them using binoculars. Daiva will always take care of him.

Prithvi is sent to Mumbai after he confesses his love to his young schoolmate Swathi, to which Swathi's father objects. Fourteen years later, it is revealed that Prithvi's brother, Ramu, is engaged in a court case against Shivaram concerning the market's demolition. Shivaram arrives at the market to warn the shopkeepers and engages in violence, at which point Prithvi returns from Mumbai. After a verbal duel, Shivaram's goons attack Prithvi, and a fight ensues. Prithvi is about to be knifed but is saved in the nick of time by his childhood friend, Pranav.

Later in the film, Prithvi is engaged to Swathi. Shivaram and his brother Babu, in an attempt to scare Prithvi back to Mumbai, try to kidnap Swathi's sister even as Swathi and Prithvi are praying at *Ajja*'s shrine. The visuals cut quickly between these two locations, and Pranav, yet again, appears at the crime scene to save Swathi's sister from Babu and his goons. Beyond these sequences, Pranav is never seen or mentioned in the film. Pranav's miraculous appearance whenever Prithvi or his loved ones are threatened is not a mere coincidence but *Ajja*'s divine intervention.

In the climax sequence of *Barsa,* Shivaram and his goons surround Prithvi in a cricket ground and attack him with an iron club, rendering him semiconscious. At this point, Prithvi has visions of *Koragajja* in his divine form – shown for the first time in the film. Encouraged by *Koragajja*'s appearance, Prithvi gains enough strength and courage to fight back and beat up the goons before falling unconscious again. The film ends with Prithvi in the hospital, surrounded by his family and friends from the market. One of the shopkeepers says:

> When you fell to the ground, I thought you won't stand again. But when you did stand, we felt alive again.

Prithvi responds, even as the camera shifts to a frontal frame, with Prithvi almost addressing the audience:

> I didn't stand up; Ajja made me stand. Ajja will never leave behind his devotees, and that is, for sure, the truth.

Even as everyone else is engaged in banter, Prithvi sees the curtains blowing. Prithvi reacts to it, indicating that Koragajja is communicating through the wind. The end credits appear with a message:

> Koragajja will watch the backs of those who believe in him; he will only shower the rain of happiness on them.[35]

Kantara dealt with *Daivaradhane* upfront; one cannot remove faith from the film. Unlike *Kantara*, *Barsa* subverts the typical action-comedy formula of Indian cinema, attributing all heroism to divine intervention.

Further, faith is not something that is merely practised to its ends; it is a means to bind people together. Faith can be used as a tool for noble causes as long as the Tuluva is a believer and is willing to be punished. Such noble causes may include keeping a family together or upholding tradition, as in *Yesa* (2017).

Yesa is the story of 'Swarna Gutthu', a *gutthumane* belonging to Vitthal Shetty, a Yakshagana performer and teacher. Shetty's daughter, Baby, elopes with her lover, who hails from Mumbai, and steals the gold ornaments of the daiva, who resides on their premises. Her elopement and the insult caused to the *daivas* by the theft result in a feud between her two brothers, who divide the household and separate their parents. Years later, Rahul, a Tulu-speaking person from Mumbai, impersonates Baby's son Abhishek and arrives at Swarna Gutthu to reunite Shetty's divided family. It is revealed that Rahul had killed Abhishek in a car accident and had decided to impersonate Abhishek to fulfil his incomplete dream of reuniting *Swarna Gutthu*.

In his quest, Rahul realizes the importance of the *daiva* to the household and leverages the family's faith in *daivas* to fulfil his quest.

He tricks them into believing that the *daiva*, who had disappeared after the family stopped worshipping following Baby's elopement, has returned to their land, signifying a positive development. However, Shetty's granddaughter, Tanu, finds out that Rahul has impersonated the daiva. She confronts him, saying:

> For you Mumbai people, these deities may be a bhoota, but for us, they are the eternal truth.

Daivas cannot be reduced to mere theatrical enjoyment of annual *Kola* festivals as enjoyed by the diasporic Tuluvas. *Daivas* are part of the material reality of the Tuluvas and must be treated with respect. However, Rahul justifies his act by comparing his use of faith towards the noble cause of uniting a feuding family to lighting a 'lamp' with a 'matchstick' only to 'illuminate' an entire household. Rahul adds:

> If it means I have insulted the daivas by using them for my personal quest to unite the family, then I'm willing to face any punishment meted out by your eternal truth.

Rahul's use of faith is pardonable, given his intent to unite the family. However, this family unification is not merely a humane act to keep relations intact.

To emphasize the ideal, the Tuluva demonstrates religiosity in day-to-day life, has unquestionable faith in the *daivas*, respects the promises made to the *daivas*, honours their prophecies and judicial rulings, and upholds faith at all costs. The ideal Tuluva condones violence only when such violence is sanctioned by the *divine*. A Tuluva can emulate the ideal Tuluva even if they are non-believers – there is always the option of recourse to faith as long as the *daivas* have not been mocked or insulted. The ideal Tuluva upholds faith to keep a family or even the nation together so that faith survives eternally through tradition.

When viewed through the sociological categories of caste, religion, gender and language, the question arises as to whom this ideal Tuluva represents. Can any Tuluva with faith in *daivas* become a Tuluva? Or are there rules of inclusion and exclusion?

The Ideal Tuluva as a Tulu-Speaking Male Bunt

While the ideal Tuluva is a figure who upholds the sacrosanct nature of *Daivaradhane* and the pantheon of *daivas* that rule over the sacred territories of Tulunad, this ideal figure tends to be an exclusionary figure when viewed through the lens of caste, religion, gender and language. The ideal Tuluva is always a Tulu-speaking male Hindu figure who hails from a non-Brahmin caste and emulates the traditions and practices of the upper-caste Bunt community to preserve them. Such an emulation requires the Tuluvas to perform their caste-based traditional roles perpetually, rendering the figure as a Bunt's view of who the ideal Tuluva is. Crucially, despite the acceptance of the *daivas* among the non-Hindus of Tulunad, the ideal Tuluva cannot be a non-Hindu – reducing their status to that of subjects.

Tuluva women have often been argued to have better social and economic status compared to women in other parts of Karnataka and India, owing to matrilineal traditions and high literacy rates, with women taking on important roles in both public and private spheres (Hegde 2019: 598). Yet, in the films analysed in this chapter, barring Manu's wife (a doctor) in *Kanchilda Baale* and Leela (a forest constable) in *Kantara*, most women characters are limited to the domestic space. The female Tuluva in films has a well-defined role with two key functions: as a carrier of the bloodline to ensure continuity of tradition, exemplified by Jaanu and the importance given to the *Kanchil* vow, and as a custodian of tradition who keeps the male Tuluva on track. In *Barsa*, Swathi's role is limited to that of Prithvi's love interest and nothing more. In *Yesa*, Tanu schools Rahul about the centrality and sanctity of *daivas* to the Tuluvas. In Kantara, Shiva's mother also ridicules him for violating his faith. In *Kanchilda Baale*, Ajjamma's promise to conduct the *Kanchil* ritual, her efforts to convince Manu to fulfil it, the efforts of Manu's wife in convincing him, the importance of Jaanu as an inheritor of wealth, tradition and family lineage as per *Aliya Santana* traditions – all point to this role of the woman as a custodian of tradition. Even though Jaanu and Ajjamma play a crucial role in *Kanchilda Baale*, the film is essentially Manu's journey of transformation, reducing Ajjamma to the role of being a custodian of tradition and Jaanu as the inheritor of that role.

Despite the matrilineal social structure among the Tuluvas, historically, Tuluva women have been structurally determined to be the custodians of tradition and responsible for the 'purity' and 'continuity' of the bloodline through self-control of caste purity and sexuality, resulting in unfavourable situations for women, as exemplified by the myth of Siri (Claus 1975, 51–3). The annual rituals associated with Siri, where Tuluva women facing social and familial issues congregate together and undergo mass possessions (Claus 1975, 47), persist even today. Increasingly, however, male control over female morality appears to be normalized as evidenced by the 2009 assault of pub-going young women in Mangaluru (Hegde 2019: 606). Violence against women, too, has increased in the region and is often attributed to 'a migrant population' and 'external influences'. The films are reflective of such realities, not as a social commentary on the issues faced by Tuluva women, but to posit the Tuluva male as the protector of women. All films focus primarily on the journey of male lead protagonists. The ideal Tuluva projected in cinema is not a female but a male. However, it is not just any male, but a male who embodies faith and religiosity and is also a protector of women.

In a market scene in *Barsa,* Shivaram assaults an elderly woman, pushing her to the ground and injuring her, in a bid to threaten the shopkeepers. At this point, the adult Prithvi is introduced. Prithvi helps the woman stand up and ensures his friend attends to her. Shivaram asks who Prithvi is, to which he replies:

> Sir, I'm a common man like everyone. My boss and I were coming from the airport and saw a roadblock outside the market. We thought something was wrong. I came inside to find a mother crying. It could be your mother or mine, but if someone tolerates a mother's tears, he cannot be a Tuluva. It's a bad scene.

Further, the ideal Tuluva is understandably proud of his language, speaks it with pride, and preserves it. Upon realizing that Prithvi is a threat, Shivaram pretends to surrender and asks Prithvi to leave, only to mock Prithvi's use of English behind his back. Prithvi returns, and Shivaram asks why he has come back. Prithvi responds:

> I may have run out of my English, but I need to save my mother tongue, Tulu, right? That's why I'm back.

This linguistic pride is present in *Kanchilda Baale,* too. On their way from the airport to Ajjamma's house, Manu casually mocks his wife and Jaanu about their English, asking them to speak in Tulu with Ajjamma. Later, Ajjamma expresses to Manu, saying, "You left the country but did not leave our language.'

The ideal Tuluva is a Tulu-speaking male who takes pride in protecting women and the Tuluva nation. In Indian film narratives, community formations such as family, village, etc., constituted by archetypal, recognizable characters, can be interpreted as allegories of the nation (Raghavendra 2021: 94). Therefore, Shiva fighting Suttur's goons to protect an abstract, fictional *Kadabettu*[36] village in *Kantara* and Prithvi fighting Shivaram's goons to protect a fictional market in *Barsa* can be interpreted as the ideal Tuluva protecting the allegorical nation.

The previous section has demonstrated that the ideal Tuluva always has faith in the *daivas*. Notably, in the world of *daivas*, Muslim *daivas* like *Bobbarya* and *Ali Bhuta* exist (Upadhyaya & Upadhyaya 2002: 114–15), and Muslims attend *Kola* of such deities. Can a Muslim, a Christian or any person from another faith be the ideal Tuluva?

Even though Christianity arrived in Tulunad in the sixteenth century with the post-inquisition migration of Konkani-speaking Roman Catholics from Goa (Pinto 2012: 997), it was the arrival of the missionaries in the late eighteenth and early nineteenth centuries that brought Christianity to native Tuluvas through conversions. In what is termed as the 'Tulu movement' (not to be confused with the political movement mentioned earlier) by the Basel mission, a large number of Tuluvas, comprised mainly lower caste Billavas, in addition to Mogaveeras, and a few upper castes, converted to Christianity to escape the prevalent social hierarchical order and in the hope of gaining education and employment (Devi 2016: 549). This Tuluva Christian, however, is missing in the Tuluva imagination as seen in films. In the films analysed in this chapter, all protagonists were not only Tulu-speaking but were invariably non-Brahmin Hindus with faith in *daivas*. Among the supporting characters, only two non-Hindu archetypal characters find a place in the allegorical Tuluva nation represented by the market in *Barsa* and the village in *Kantara*.

In *Barsa*, one of the many shopkeepers in the market is a Konkani-speaking Christian named Richard Pinto, who also speaks Tulu. This simultaneous inclusion of a Konkani Christian and the absence of Tulu Christians reinforces the argument that the ideal Tuluva is always a Hindu. On the one hand, the nature of the inclusion of the Konkani Christian relegates them to a subject position, that is, they have a place in the nation but are not considered ideal. On the other hand, like the Tuluva Christians, other populous communities, including Muslims, are found largely missing in the films analysed. *Barsa*, *Yesa* and *Kanchilda Baale* do not have Muslim characters. Only in *Kantara* does one spot a Muslim in the form of Abbu, who bursts *garnaal*, a traditional firecracker yearly during the *Kola*. Indeed, it is true that it is primarily Muslims who burst the gernel at the *Kola* events, typically outside the main premises where the ceremony takes place. *Kantara* even has its own 'good Muslim' versus 'bad Muslim' moment when, in the climax, Abbu is surrounded by Muslim goons of Suttur who, on finding Abbu, greet him with a 'Salaam Wal-e-kum', to which Abbu responds with a 'Wal-e-kum Salaam', bursting a grenade and blowing up the house. The good Muslim has a place in the Tuluva nation as a subject as long as he upholds the land and its traditions. Beyond Abbu in *Kantara*, none of the films has an identifiable Muslim character.[37]

The ideal Tuluva is, therefore, an exclusionary figure based on gender, language and religion. Those who fall outside this ideal have their place as subjects, as long as they do not question the faith central to the nation, that is, *Daivaradhane*. Who is the ideal Tuluva when seen through the lens of caste?

Tuluva society, like elsewhere in India, is based on a caste-based hierarchical social order, with Brahmins and Bunts being the traditional landholders and rulers, while Billavas and Mogaveeras, considered the 'backbone' of Tulu culture, particularly in the context of *Daivaradhane*, were engaged in agriculture and fishing, respectively. Billavas, particularly, shared a servant–landlord relationship with Bunts and were crucial in the faith system of *Daivaradhane* as the custodians of the *Daivas* and *Daivasthanas* patronized by the Bunts (Koudur 2020: 253–4). The mass conversion of Billavas to Christianity disturbed the traditional social order and their associated relationships, leading to backlash and even 'appropriation' of the lower caste language and culture by the upper castes (Koudur 2020: 261). Such power struggles continued into the postcolonial era, with the upper castes striving to deny, or at least delay, full ownership of land by the lower castes, including Billavas, in the context of land reforms in the 1970s that triggered mass migrations of Bunts and Brahmins (Damle 1989: 84–90). In the contemporary context, caste-based organizations and caste-based surnames continue to visibly exist. Tulu films, however, paint a different picture of the Tuluva nation and, thereby, the ideal Tuluva.

The ideal Tuluva cannot be a Brahmin. Barring Kuppanna Bhat in *Kanchilda Baale*, none of the films analysed had any consequential Brahmin character, despite their sizeable population and influence in the region. To recall, Kuppanna Bhat is portrayed as an oppressor and a carrier of superstition from which the Tuluvas must distance themselves. Beyond the Brahmin in *Kanchilda Baale*, the caste identities of the protagonists and important characters are either absent or obfuscated.

From a linguistic perspective, even though Tulu has many caste and region-specific dialects, only the Brahmins and specific tribal and Dalit castes use caste-specific dialects. Such use of caste-specific dialects in either media or daily communication has increasingly limited itself to domestic or intra-caste social interactions, with the specific northern dialect spoken in the Mangaluru region occupying the dominant space as the lingua franca as well as the language of Tulu media (Shetty 2008: 64). In line with this social reality, almost all characters in the films, not to mention the protagonists, speak this dominant dialect, thereby rendering the caste identity of the characters ambiguous. This is further exacerbated by the absence of characters' surnames, despite caste-specific surnames being the norm in Tulunad. Barring *Barsa*, where a comical supporting character is a priest of a *daivasthana* and can be interpreted as a *Billava-Poojari* based on traditional caste functions, most Tuluvas projected in the films cannot be identified by their caste functions given their secularized presentations. However, the Dalit is instantly recognizable through representation in both *Kantara* and *Kanchilda Baale*.

Shiva is a Dalit in Kantara, given that *daivanarthakas* always belong to specific Dalit sub-castes (Upadhyaya & Upadhyaya 2002: 24). *Kantara* critiques untouchability in many scenes – from how and where Shiva and his accomplices are served food, to their being restricted from entering Suttur's house, to Guruva's hesitancy to board Suttur's vehicle, to Guruva's surprise at being touched by Suttur, to Suttur's act of entering Guruva's house, which surprises all the villagers, to Suttur washing his hands after touching Shiva before sending him off to kill the forest officer. Evidently, the critique of untouchability is strong and not merely nominal. Against these repeated depictions of untouchability, Shiva's act of entering Suttur's house and dining with him just before the climax can be read as not just transcending caste boundaries but as an act of emancipation. The idea of emancipation becomes more potent in the climax when Shiva finally kills Suttur. To that extent, *Kantara* is indeed a critique of untouchability and can be viewed as Dalit emancipation and as the uprising of the Dalit against the upper castes. However, an alternative reading of *Kantara*'s portrayal of Dalits is possible.

In the climax of *Kantara*, Suttur and his goons arrive at the village to burn it down. A violent battle ensues between Suttur's goons and the villagers, who are led by Shiva, Muralidhar and the forest guards. The battle intensifies, and Shiva is overpowered, strangled and thrown violently against a stone. Shiva bleeds and falls unconscious. Suttur announces Shiva's death and suggests that the villagers evacuate. He asks Shiva's mother and other widows to become his consorts. At this point, even as the visuals show an unconscious Shiva, the voice of *Panjurli Daiva* from the prologue is heard, warning the king and prophesizing the arrival of 'the ferocious *Guliga*' to protect *dharma* upon the violation of the promise in the future. The stone against which Shiva is thrown is revealed to be that of *Guliga*, marked by the symbol of a *Trishula*, pointing to the truth of the prophecy. Shiva's father appears as *Panjurli* and roars to awaken the *Guliga* in Shiva. Now possessed by *Guliga*, Shiva fights Suttur's goons, dancing violently and displaying superhuman strength and abilities with no signs of struggle or weakness. The villagers recognize Shiva's possession by *Guliga* and surround him with their hands folded in devotion. They provide a bag of puffed rice as an offering to *Guliga,* who accepts and savours it, even as he tackles and kills all of Suttur's men. Approaching Suttur with a sword in his hand, Guliga announces, just before he kills Suttur:

> Carrying out your atrocities in the land of Swami Varahamurthy, did you forget the protector of the land who guards its borders from those who violate *dharma*? My lord *Panjurli* might forgive, but not me. By offering your blood to this soil in return for your sins, I shall save *dharma.*

The invoking of *Panjurli*'s prophecy about *Guliga*'s arrival to protect *dharma,* coupled with the villagers praying to *Guliga* and providing him offerings, establishes that Guliga, not Shiva, wields the sword to kill Suttur. As mentioned earlier, the *daivanarthaka,* when possessed, is a *daiva*. To delineate one from the other is a violation of faith. *Kantara*'s turn to divine intervention to serve justice over the human being signifies that Shiva, as a Dalit, can violate caste boundaries and question power but cannot serve justice. Only *Guliga* – the deity – can.

As *Kantara* ends, Shiva returns not just to faith but to his hereditary function of being a *daivanarthaka,* suggesting an eternal continuity of traditional roles that preserve faith and tradition. When Shiva, in the form of *Panjurli*, disappears into the forest to unite with his father, and his wife Leela touches her belly, it signifies the new life growing in her womb, ready for its hereditary role. An alternative reading of *Kantara* would be to argue that even as *Kantara* speaks up against untouchability, the question of Dalit emancipation remains within the boundaries of faith and does not transcend it. The real-life *daivanarthakas* must stay within the fold of faith and continue their hereditary role for the survival of the faith itself.[38] By doing so, *Kantara* ends up preserving present caste-based practices.

Such appeals to preserve faith and tradition by upholding caste-based practices are present in *Kanchilda Baale,* too. In one scene, Jaanu serves Koraga's grandson Kaalu food in the backyard of the main house. Intrigued, Jaanu asks why he is served food in the backyard, to which Ajjamma says, 'He likes sitting in the open air in the backyard when eating'. Kaalu finishes eating and cleans the place with water. Jaanu questions Ajjamma about this, and she responds, 'To purify the place where he ate so that nobody steps foot on the place where he ate.' While *Kanchilda Baale* reinforces notions of purity and pollution, *Kantara* speaks against untouchability while upholding caste-based practices and professions. The Dalit is included in the Tuluva nation, but the Dalit performs their hereditary role to ensure the survival of faith. It is now established that, barring the Brahmin, who is a threat to faith, and the Dalit, whose hereditary role is foundational to faith, caste is generally obfuscated in the selected films. However, *Yesa*, which is the story of reuniting the divided family of Vitthal Shetty – a Bunt, set in the 'Swarna Gutthu' *gutthumane*, is the sole exception as it explicitly mentions the 'Shetty' surname and depicts a *gutthumane*, indicating that the story is of a Bunt family. A closer analysis of other films reveals the Bunt-centrism of all films.

Indian cinema has typically relied on 'surnameless and, thus, regionless, casteless, ethnically non-identifiable' archetypal characters, which seem universal but carry enough 'visual codes' that help identify their particular identities (Chakravarty 1993: 200–1). Apart from the Brahmin, that is, Kuppanna Bhat, the Tuluva was identifiable by surname only when the cinematic narrative revolved around a particular family's internal issues. In all other cases, the Tuluva was a surnameless person, not belonging to any particular town or city in Tulunad or explicitly identifiable with any particular caste. However, it is through the 'visual code' that appears in the form of the *gutthumane* that one can identify the Bunt-ness of the Tuluva and the Bunt-centric view of all films.

The *gutthumane* is a 'traditional Bunt manor house'. It typically houses a *daivasthana* for the *daiva* within the premises; larger ones may dedicate an 'operational space' to host *Kola* (Benjamin 2017: 206). The *gutthumane* acts as a site where the cinematic narrative unfolds in *Barsa*, *Kanchilda Baale*, *Yesa* and *Kantara,* showcasing the Bunt-ness of those who occupy the space. Like *Yesa*, the protagonists' families in *Barsa* and *Kanchilda Baale* live in *gutthumanes,* identifiable by their grandeur, architectural style and the presence of a *daivasthana*. In *Kanchilda Baale*, the Bunt-ness is also evident through multiple social and cultural markers, including vast agricultural land ownership and feudal-style relationships with the farm labourers. Most importantly, Mahabala explains to Manu's wife that they trace their ancestry to the royal family of Barkur and have consistently practised *Aliya Santana,* a typical Bunt practice. In both *Yesa* and *Kantara*, *Aliya Santana* is violated by the demand for a patrilineal property division. In *Yesa*, the *daiva* becomes a binding force to ensure this tradition is upheld and property remains intact. In *Kantara,* the *daiva* punishes those who violate the tradition so that the land remains intact.

Crucially, for the Bunts, *Daivaradhane* and *Aliya Santana* are closely linked and quintessentially tied to land and property more than others, given their history as a feudal landholding community. The *Daivas* who reside on their premises and in their lands are akin to holding 'title' to the property, bestowing them with 'social position and honour' (Benjamin 2017: 206). *Aliya Santana* protected lands from fragmentation through matrilineal inheritance. Sometimes followed by other landholders, changes brought to this traditional law, coupled with changes to tenancy rules, mainly affected the Bunts, who, given their historical domination in land ownership, lost large tracts of land and much of their economic power, triggering en-masse migration (Benjamin 2017, 216). *Yesa*, *Kantara* and to some extent, *Kanchilda Baale*, reflect the community's anxieties about land fragmentation. Notably, *Kantara* depicts three Bunt landlords who uphold the tradition: the king from 1847, the landlord from 1970 and Rajeeva, the custodian of *Panjurli's bhandara* from 1990, subtly providing the Bunt with ideals they can model themselves on – telling him what to be, what not to be and the consequences of behaviours of the latter kind. Other films, however, address the Bunt on a positive note. Rahul, Manu and Prithvi embody their Bunt-ness as much as they uphold linguistic pride and faith. They are kind-hearted men at the nation's service – be it by protecting innocent shopkeepers from greedy developers like Prithvi does or building hospitals and protecting the oppressed from oppressors like Manu or uniting someone's family like Rahul – the Bunt is omnipresent in all the films. Does that make the Bunt an ideal Tuluva?

The ideal Tuluva projected in most films cannot be identified merely through visual codes of caste, despite the visible Bunt-ness of the protagonists, as established earlier. The selected films appear to address a broader set of communities beyond the Bunts. As mentioned at the beginning of this chapter, in non-Western societies, the national ideal that enables citizenship has typically been based on the ideals of the dominant community. In the Tuluva context, therefore, the ideal Tuluva is not necessarily a Bunt. However, this chapter argues that this ideal Tuluva is a figure shaped by Bunt ideals, much like how the upper-caste Hindu ideals 'invisibly marked' the secular, normative Indian identity (Bhrugubanda, 2018: 5). This ideal equates the Tuluvaness of a Tuluva with values that the dominant community thinks are necessary for the survival of the Tuluva nation.

After land reforms, the Bunts migrated to other cities and countries. They thrived in various fields like transportation, films, entertainment, restaurants, real estate and even the underworld, arresting their slide in political and economic power.[39] The films reflect their current concerns: preventing further land fragmentation in Tulunad, particularly by the non-Tuluvas. Bunts and all landholding Tuluva castes can choose matrilineal traditions over patrilineal laws and prevent such fragmentation of land, which also fragments the jurisdiction of the *daivas*, reducing the influence of the Bunts in Tuluva society – politically and ritually. Despite fragmentation, the Bunts have enjoyed the highest ritual position in *Kola,*[40] even in lands they no longer own, owing to their ancestral identification with a particular *daiva*. This ritual power bestows them with honour and helps to maintain a dominant position, providing them with the 'prestige to control the political and moral order' of Tuluva society (Carrin 2008: 181). For faith to survive, tradition must continue 'as is', including *Aliya Santana* and the *Kola* rituals. To preserve the land from further fragmentation, non-Bunt castes must follow *Aliya Santana*. To preserve the ritual position of honour of Bunts, all castes, particularly the *daivanarthakas* who play 'the most important role in the worship' of conveying 'the divine message' (Upadhyaya & Upadhyaya 2002: 89), must follow their hereditary role and the piety that the role warrants. To step out of faith and tradition is to hand over ritual power and dominant position in the Tuluva society.

Films addressing the Tuluva identity reflect the Bunt worldview by foregrounding their vision of Tuluva society – a society where Tuluva lands remain among the Tuluvas, achieved by providing the Tuluva castes with a reminder of their tradition of *Aliya Santana* and the preservation of their faith that upholds this tradition, that is, *Daivaradhane*. The Tulu films privilege the male Tuluva, who demonstrates religiosity, ensures the continuity of tradition and upholds faith. The ideal Tuluva could be any Tulu-speaking Hindu male from a Tuluva caste; the only requirement is to stay within the fold of faith as a close-knit Tuluva nation and to protect tradition, faith and resources from the non-Tuluvas, particularly those who do not take part in the system of faith despite having the opportunity to do so. The foregrounding of Bunt practices as the ideal in all films, either directly or indirectly, establishes the ideal Tuluva as a product of the dominant Bunt view of the traditional, feudal social order to preserve the community's ritual power and their faith. Tuluvas and non-Tuluvas who do not fall within the definition of these ideals are relegated to subject positions irrespective of their caste, religion, language or gender. They all have their place in the nation as long as they subject themselves to the true faith, making it sacrosanct and inviolable. The Tuluva nation imagined through films is a Bunt view of what the nation should be.

Conclusion

Tulu sub-nationalism is a nascent movement. In its quest for official recognition of the Tulu language and statehood for Tulunad, the movement is pitching the Tuluva identity in opposition to an overarching Kannadiga identity. True to its claims of cultural syncretism, Tulunad can be imagined as a multi-cultural, multi-lingual nation based on equality and secularity. However, the Tulu movement seeks to privilege the Tulu language and imagine a nation of Tulu speakers. However, this presumed linguistic Tuluva identity collapses into an ethnic identity, as this chapter demonstrates through films. The Tuluva identity foregrounds an exclusive ideal Tuluva, which subjectivizes non-Tulu and Tulu speakers who belong to faiths and castes outside the region. The inviolable nature of *Daivaradhane,* represented by foregrounding this religiosity, provides a subject position within the faith as long as it is upheld. The ideal Tuluva imagined in the films analysed in this chapter is a Tulu-speaking male who demonstrates religiosity and pride in language and preserves and practices Tuluva cultures and traditions. The centrality of the *daivas* and *Aliya Santana* to the Tuluva identity turns it into an exclusionary figure by assigning subject positions to anyone who does not meet the ideal. This ideal mirrors the Bunt view of the society, aiding in their preservation of ritual power and political domination. If deployed in the sub-nationalist Tulu movement, the Tuluva identity becomes exclusionary based on language, religion, caste and gender and imagines a nation that privileges the ideology of the dominant caste even if a cohesive society is achieved in the form of a syncretic system of *Daivaradhane.*

Filmography

- ***Barsa*** (2016) [Film], Dir. Devadas Kapikad, India: Bolli Movies.
- ***Gaggara*** (2008) [Film], Dir. Shivadhwaj Shetty, India.
- ***Garuda Gamana Vrushabha Vahana*** (2021) [Film], Dir. Raj B. Shetty, India: KRG Studios.
- ***Kaljiga*** (2024) [Film], Dir. Devadas Kapikad, India: Bolli Movies.
- ***Kanchilda Baale*** (2011) [Film], Dir. Raghunath Rai Kumbra, India: Kaala Bhairava Creations.
- ***Kantara*** (2022) [Film], Dir. Rishab Shetty, India: Hombale Films.
- ***Koti Chennayya*** (1973) [Film], Dir. Muddu Suvarna, India: Praja Films.
- ***Koti Chennayya*** (2007) [Film], Dir. Anand P. Raju, India.
- ***Paddayi*** (2017) [Film], Dir. Abhaya Simha, India: Gopal Pai. Films.
- ***Pancharangi*** (2010) [Film], Dir. Yogaraj Bhat, India: Yogaraj Bhat Films.

- ***Ulidavaru Kandanthe*** (2014) [Film], Dir. Rakshit Shetty, India: Suvin Cinemas.
- ***Yesa*** (2017) [Film], Dir. M. N. Jayanth, India: U2 Cinema Talkies.

Notes

1 As per 2011 Census Data, two million speakers identify Tulu as their mother tongue, 1.5 million of them live in Dakshina Kannada and Udupi districts of Karnataka. The number may be marginally higher when bilingual Tulu speakers are included.

2 Trans. Tulu Mother.

3 See Tuluvere Paksha seeks separate statehood for Tulu Nadu, language (https://web.archive.org/web/20240524013121/https://www.deccanherald.com/india/karnataka/tuluvere-paksha-seeks-separate-statehood-for-tulu-nadu-language-967521.html).

4 See website of Tulunad Rakshana Vedike (https://tulunadurakshanavedike.blogspot.com/). Also see: Tulunadu Rakshana Vedike demands statehood (https://web.archive.org/web/20170505175039/https:/www.thehindu.com/news/national/karnataka/Tulunadu-Rakshana-Vedike-demands-statehood/article16087168.ece).

5 See website of TuluWorld (https://tuluworld.org.in/about).

6 Originally made in Kannada, *Kantara* was later dubbed in Hindi, Tamil and Telugu. Typically, Kannada films are not dubbed into Tulu as people in Tulunad speak and understand Kannada. However, the Kannada language and films are alien to those Tuluvas who grew up outside Karnataka. The Tulu version was released only after a popular demand.

7 See Kantara to be dubbed in Tulu | Kannada Movie News – *Times of India* (https://web.archive.org/web/20221030124432/https:/timesofindia.indiatimes.com/entertainment/kannada/movies/news/kantara-to-be-dubbed-in-tulu/articleshow/94856289.cms).

8 *Daivas* or *bhutas* are spirit deities worshipped in the form of stones in the faith system of *Daivaradhane,* also referred to as *Bhutaradhane* or Bhuta Worship – tradition prevalent in Tulunad. Similar to *Theyyam*

tradition of northern Kerala, it is an ancient tradition followed mostly by present-day non-Brahmin Hindus. Today, it is not uncommon for Konkanis, Brahmins or even Muslims showing faith in *daivas*. The use of the term *daivas* vis-à-vis *bhutas* has its own embedded politics with the former considered a sanskritized term, even as the term *bhuta* itself was a sanskritized term which replaced an earlier Tulu term *pey* or *peyi* (Koudur 2020: 258). This chapter uses the term *daiva* instead of *bhuta* as most films analysed use the same. For more, see Upadhyaya & Upadhyaya (2002).

9 Kannada-speaking people from Kundapura region of Tulunad follow *Daivaradhane* and social structure of Tulunad. The assertion of *Daivaradhane* as a culture associated with Tulu-language solely has its own embedded politics and is outside the scope of this chapter.

10 Paniyadi (2007: 140) argued that the Basel mission's choice to rename itself 'Karnataka Mission' led people in Tulunad to switch to Kannada domestically. For the role of Basel mission in deploying Kannada over Tulu for its operations, and the role of the upper castes of coastal Karnataka in preventing access to English education for lower castes like Billavas and Mogaveeras, and the eventual appropriation of Tulu language and practices of the lower castes by the upper castes, see Koudur (2020).

11 In *Rayara Sose* (1957) two servants speaking the Mangaluru and Belagavi Kannada dialects are positioned hierarchically lower in the houshehold. Comedy is derived from dialect differences (Raghavendra 2011: 13).

12 In *Pancharangi* (2010), the female lead's household consists of her aunt, uncle and cousin, and the uncle speaks Mangaluru Kannada. The uncle is married to a Bengalurean woman. The daughter and the niece largely speak Bengaluru dialect. Like *Rayara Sose* (1957), such characterizations aid in comical situations.

13 The household has two servants – a brother–sister pair. The brother is hardworking but not shrewd, setting up a few comical sequences. The sister, is outspoken and shrewd. She speaks the Kundapura dialect mixed with some Tulu words, creating ambiguities regarding her identity. Nevertheless, both characters occupy subordinate positions in the narrative.

14 Rakshit Shetty, Rishab Shetty and Raj Shetty are actors and filmmakers who work primarily in the Kannada film industry. Hailing from Udupi, Kundapura and Mangaluru, respectively, they have made highly popular Kannada films focusing on the culture of Tulunadu. See 'Meet the 3 Shettys Changing Kannada Cinema' – *The Hindu* (https://web.archive.org/web/20230823113928/https:/www.thehindu.com/entertainment/movies/rakshit-rishab-and-raj-meet-the-three-shettys-changing-kannada-cinema/article38112176.ece).

15 Kannada films typically use the standardized Kannada which largely derives from Mysuru region's Kannada. Increasingly, however, Bengaluru's street dialect has taken over, often in its vulgar form (Raghavendra 2011: 113).

16 Kundagannada is a dialect of Kannada spoken in Kundapura region. Even as people of the Kundapura region share almost all traditional practices of Tulunadu, they do not speak Tulu.

17 Jumaadi is an androgynous *daiva* worshipped across Tulunad and has many forms (Upadhyaya & Upadhyaya 2002: 5). Jumaadi is considered the son of Mother Goddess or the Hindu goddess Parvati (Upadhyaya & Upadhyaya 2002: 101).

18 *Banta*, or *Bante*, is a jester-like figure who accompanies *Jumaadi* (Upadhyaya & Upadhyaya 2002: 5). According to the Maayinda Maani *paddana*, *Banta* is the son of *Maayinda Mani* – a female who was transformed into spirit by Jumaadi when *Maayinda Mani*'s uncle Nagayya refuses to contribute to *Jumaadi*'s *Kola* ceremony organized by Pangulu Bannar, the local king of Kalya (Upadhyaya & Upadhyaya 2002: 56).

19 See 8.

20 Trans. ethics.

21 Trans. virtue.

22 See 'Cooku with Comali Pugazh Hurts the Sentiments of Karnataka People with His Controversial Panchuruli Look?' – *Filmibeat* (https://web.archive.org/web/20230320003219/https:/www.filmibeat.com/tamil/news/2023/cooku-with-comali-pugazh-hurts-the-sentiments-of-karnataka-people-with-his-controversial-panchuruli-348236.html). Also see 'Make Up Artist Shwetha Reddy Asks Apology, for Imitating Panjurli Daiva' – *Kannada Filmibeat* (https://web.archive.org/web/20240512075909/https:/kannada.filmibeat.com/news/make-up-artist-shwetha-reddy-asks-apology-for-imitating-panjurli-daiva-062813.html).

23 'Koragajja Nema' ritual cancelled in Bengaluru after backlash from Tulu people: Report (https://www.hindustantimes.com/cities/bengaluru-news/koragajja-nema-ritual-cancelled-in-bengaluru-after-backlash-from-tulus-101669360930961.html).

24 The Kannadiga identity typically refers to Kannada speakers. It is also an umbrella term used for people of Karnataka in popular parlance.

25 See Darshan (2016) and Darshan (2017) for *Kanchilda Baale* and *Gaggara*, respectively.

26 Can a handful of filmmakers be called an industry? In Tulu, the answer is yes (https://web.archive.org/web/20150515211741/https:/scroll.in/article/718862/can-a-handful-of-filmmakers-be-called-an-industry-in-tulu-the-answer-is-yes). Also see, 'After 50 Years, Tulu Cinema Needs a New Lease of Life' – *The Week* (https://web.archive.org/web/20231031232924/https:/www.theweek.in/theweek/leisure/2022/03/19/after-50-years-tulu-cinema-needs-a-new-lease-of-life.html).

27 'How Tulu Cinema Came out of Shadows to Script Identity' | Latest News India – *Hindustan Times* (https://web.archive.org/web/20230523052222/https:/www.hindustantimes.com/india-news/how-tulu-cinema-came-out-of-shadows-to-script-identity-101628967185909.html).

28 *Panjurli* is considered one of the most powerful *daivas* and is worshipped across Tulunad in many forms. The *daiva* is of totemistic origin and is associated to the wild boar, making wild boars sacred in Tulunad. *Panjurli* primarily resides in agricultural and forest lands.

29 *Guliga* – the custodian of night, the controller of movement of ghosts, is a 'blood sucking', ferocious *daiva* who can cause great harm on being dissatisfied (Upadhyaya & Upadhyaya 2002: 116). In *Kantara*'s prologue, *Guliga* is introduced as the custodian of faith who will descend if faith and the land *Panjurli* resides are at threat.

30 *Bhuta Kola*, or simply *Kola*, is a periodically organized festival where a *daivanarthaka* impersonates, and is possessed by, the *daiva* for whom the ritual is being organized. The *Kola* is performed on a fixed day of the year, or when an astrologer or priest says it must be conducted (Upadhyaya & Upadhyaya 2002: 23). The *Kola* is an occasion for concerned *daivas* to listen to their devotees, provide and solutions to personal and social issues including legal and judicial problems (Upadhyaya & Upadhyaya 2002: 22).

31 Shiva hunting wild boars is, therefore, an insult to the faith, and the very deity who protects *Kadabettu* and who his father impersonated during the *Kola*.

32 In the film the character's name is not mentioned and is simply referred to as Dani (trans. Landlord).

33 A Dalit sub-caste.

34 Moola nambiken oripule, mooDha nambiken danTle.

35 'Nambinaklen beri buDayi Koragajje, imbuda 'Barsa'ne barpaave'.

36 There is a village called Kadabettu located closer to Mangaluru in Tulunad. In *Kantara*, the Kambala sequence is shot in Keradi and there is no village named Kadabettu in its surroundings.

37 This general absence of Muslims in the films and the projection of a particular kind of Muslim as the desirable subject seem to reflect the increasing tensions between the Hindus and Muslims in the region. See 'Symbols of Unity Now Ground for Religious Divide' | Latest News India – *Hindustan Times* (https://web.archive.org/web/20220610194450/https:/www.hindustantimes.com/india-news/symbols-of-unity-now-ground-for-religious-divide-101649011277248.html).

38 Such portrayal has been criticized by some Dalits and even raised questions of upper-caste appropriation of the Dalit's role in the ritual.

39 For more, see Benjamin (2017).

40 With rare exceptions when a Brahmin might occupy a higher symbolic position without much involvement.

References

Benjamin, Solomon. 2017. 'Multilayered Urbanization of the South Canara Territory'. In *Subaltern Urbanisation in India. Exploring Urban Change in South Asia*, edited by E. Denis and M. H. Zérah, 199–233. New Delhi: Springer.

Bhrugubanda, Uma Maheswari. 2018. *Deities and Devotees: Cinema, Religion, and Politics in South India*. New Delhi: Oxford University Press.

Caldwell, Robert. 1875. *A Comparative Grammar of the Dravidian or South-Indian Family of Languages*. London: Trübner.

Carrin, Marine. 2008. 'Dialectics of Autochthony and Foreignness in the Bhuta Cult of the Kanara Coast of the Foreigner in Karnataka'. In *Journeys and Dwellings Indian Ocean Themes in South Asia,* edited by E. Simpson and H. Basu, 174–97. Hyderabad: Orient Longman.

Chakravarty, Sumita S. 1993. *National Identity in Indian Popular Cinema, 1947–1987*. Austin, TX: University of Texas Press.

Claus, Peter J. 1975. 'The Siri Myth and Ritual: A Mass Possession Cult of South India'. *Ethnology* 14(1): 47–58.

Damle, Chandrashekhara Bhat. 1989. 'Impact of Tenancy Legislation and Changing Agrarian Relations: A Case of Dakshina Kannada District, Karnataka'. *Social Scientist* 17(11/12): 83–97.

Darshan, B. M. (2016). 'Ethnography and Tulu Cinema: A Case Study of the film "Kanchilda Baale"'. *International Education and Research Journal* 2(8): 68–9.

Darshan, B. M. (2017). 'Ethnography in Tulu Films: An Analysis of the Film "Gaggara"'. *Global Journal for Research Analysis* 6(4): 612–13.

Darshan, B. M. and Ramya K. Prasad. 2020. 'Cinema in Formation of Cultural Identities: A Study on the Ethnographic Films and People's Perception on Tulu Films'. *International Journal of Psychosocial Rehabilitation* 24(8): 9256–67.

Dechamma, Sowmya. 2010. 'Is There a Kodava Cinema?' In *Cinemas of South India,* edited by S. Dechamma and E. S. Prakash, 190–216. New Delhi: Oxford University Press.

Devi B., Meera. 2016. 'The Billavas and the Socio-religious Movements in the Colonial South Kanara'. *Proceedings of the Indian History Congress* 77: 546–53.

Gellner, Ernest. 1983. *Nations and Nationalism*. Oxford: Blackwell.

Grotenhuis, Rene. 2016. 'Nation-Building: Identity and Identification, Process and Content'. In *Nation-Building as Necessary Effort in Fragile States*, 109–24. Amsterdam: Amsterdam University Press.

Gururaj, P., Abhinandan Kulal, N. Abhishek and Swarn G. Kanchan. 2022. 'Student's Attitude and Interest towards Learning Tulu Script'. *International Journal of Research – Granthaalaya* 10(8): 174–87.

Hegde, Shakila. 2019. 'Issues and Concerns of Women in India with Special Reference to Women in Dakshina Kannada District of Karnataka'. *International Journal of Emerging Technologies and Innovative Research* 6(6): 597–608.

Koudur, Shashikantha. 2020. 'Languages, Castes and Hierarchy: Basel Mission in Nineteenth Century Coastal Karnataka'. *South Asia Research* 40(2): 250–65.

Mangaluru, H. 2007. 'Chalanachitra'. In *Tulu Saahitya Charitre,* edited by B. V. Rai et al., 990–9. Hampi: Prasaranga, Kannada University.

Paniyadi, Srinivasa Upadhyaya. 2007. 'Tulunaad Prantada Bedike: Ondu Samarthane'. In *Tulu Saahitya Charitre,* edited by B. V. Rai et al., 139–40. Hampi: Prasaranga, Kannada University.

Pinto, Vishanz. 2012. 'Theocratic Hegemony of the Roman Catholic Church and the Response of the Faithful – The Case of the Mangalore Diocese'. *Proceedings of the Indian History Congress* 73: 997–1005.

Raghavendra, M. K. 2011. *Bipolar Identity: Region, Nation, and the Kannada Language Film*. India: OUP India.

Raghavendra, M. K. 2017. *Beyond Bollywood: The Cinemas of South India.* India: HarperCollins Publishers India.

Raghavendra, M. K. 2021. *Philosophical Issues in Indian Cinema: Approximate Terms and Concepts.* United Kingdom: Routledge.

Schlesinger, Philip. 2000. 'The Sociological Scope of "National Cinema"'. In *Cinema and Nation,* edited by M. Hjort and S. Mackenzie, 19–31. New York: Routledge.

Shetty, Malavika Leeladhar. 2008. *Television and the Construction of Tulu Identity in South India.* Austin, TX: The University of Texas at Austin.

Someshwar, Amrith, Muraleedhara Upadhya Hiriadka and B. R. Nagesh. 2000. 'Amrith Someshwar in Conversation with Muraleedhara Upadhya Hiriadka'. *Indian Literature* 44, no. 4 (198): 81–96.

Sorlin, Pierre. 1996. *Italian National Cinema.* London: Routledge.

Uberoi, Patricia. 1998. 'The Diaspora Comes Home: Disciplining Desire in DDLJ'. *Contributions to Indian Sociology* 32(2): 305–36.

Upadhyaya, Uliyar Padmanabha and Susheela Padmanabha Upadhyaya. 2002. *Folk Rituals (of the Tuluva Region of Coastal Karnataka).* Udupi: Regional Resource Centre for Folk Performing Arts, MGM College.

Zake, Ieva. 2002. 'The Construction of National(ist) Subject: Applying the Ideas of Louis Althusser and Michel Foucault to Nationalism'. *Social Thought & Research* 25(1/2): 217–46.

4

More than Trees: Curated Forests and Unruly Narratives of Place

Jayaditya Vittal and Gayathri Prabhu

Introduction

The Saalumarada Thimmakka Tree Park in Manipal, one among nearly 124 parks set up across the state of Karnataka by the State Forest Department to provide 'urban dwellers an opportunity to experience nature at accessible distance' (aranya.gov.in), invites scrutiny of curious contradictions between overlapping eco-cultural narratives. This particular location, of about 3000 sq metres, hemmed in by fences and stone walls, paradoxically sits in the midst of a wooded stretch of the Western Ghats that is sparsely populated, patchily cultivated and already within a larger forested region. The demarcated space with entry timings and entry fees features an elaborate layout designed with the express intent of curation (primarily vegetation), of education (through sculptural displays and signages), and of leisure and entertainment (such as tree houses and play areas for children as well as an attached canteen) that cater to a steady stream of daily visitors, mostly from nearby towns. The curated forest area thus speaks of state intent towards conservation practices within broader environmental discourses, and equally represents the embodied experiences of its visitors. This study combines observational fieldwork and textual analysis to argue for the centrality and necessary ambiguities of narratives – drawn from iconic personalities, local lore, policy directives, sustainability rhetoric or personal impressions – in the manifestations of place-based practices and environmental citizenship. Such narratives, we argue, tend to be unruly, not amenable to neat categorizations or conclusive outcomes, and in fact often function as strategically disruptive. The adjective 'unruly' has been repeatedly favoured by several eco-scholars and writers of environmental literature in recent years often to delineate the autonomy of biotic and abiotic processes – there is now an extensive scholarship on this (Amrith 2018: 5–10). Species and environments have thus been considered 'unruly' to describe modes of existence that disrupt human and State expectations of them, leading to unintended and not envisioned implications and outcomes. This chapter draws on Peter J Taylor's conception of 'unruly complexity' as any 'ongoing change in the structure of situations that have built up over time from heterogeneous components and are embedded or situated within wider dynamics' (Taylor 2005: xvii). The heterogeneity becomes indispensable and crucial in the rhetoric of spaces reserved for public edification. A park-designated area, such as the Saalumarada Thimmakka Tree Park in Manipal, located deep within one of the world's eight hottest biodiversity hotspots (Centre, UNESCO World Heritage, 'Maps') where walled boundaries cannot either contain or bracket out rich biotic and abiotic elements, brings to the fore instructive tensions between curation and unruliness in the context of the ecologically sensitive mountain ranges of the Western Ghats.

The Karnataka Forest Department launched the scheme of setting up Tree Parks in 2011–12 with an express intent of planting local species in areas not far from towns and by 'providing chain-link fence of stone/brick wall so that the planted seedlings would get maximum protection from biotic interference' ('Saalumarada Thimmakka Vruksha Udyana Yojane', Karnataka Forest Department, 2012). The scheme was renamed after Thimmakka, a grassroots environmentalist, in 2017–18. The forest department's identification of a threat ('biotic interference') and the offering of a solution (fencing) with the intention of promoting proximity between people and 'nature' displays an anxiety around bringing together conservation, curation and being relatable as well as accessible to public interest or patronage. In Manipal, the area reserved for the Tree Park is situated on the outskirts of a campus town, which in turn is a fairly seamless adjacency to the district headquarters of Udupi in coastal Karnataka. Perched atop a hill, with one flank of the tree park sharing its boundary with the sprawling campus of a well-known business school, its other sides are surrounded by swathes of open forests. The sporadic connectivity of affordable public transport to this location necessitates the arrangement of other modes of transport such as auto rickshaws, school buses or private vehicles to access the Tree Park. This tends to preclude wider public participation in the space and has an influence upon the socio-economic profile[1] of the visiting public. Visitors are greeted at the entrance of the park by a gigantic mural of an older woman's smiling face alongside a signage that explicates Thimmakka's environmental activism. Immediately past the park's large metal gates is a central sculpture of a tree that takes on the shape of a human with joined palms. This anthropomorphic tree-sculpture cast in cement, an incongruous not-natural element in the context of trees, is only the first of many such large sculptures of flora and fauna speckled across the precincts of the park. As with the sculptures, the notion and narrative of the 'Tree' function as a discursive synecdoche as much for forests as for attendant eco-cultural narratives in addressing an assumed expectation of infotainment by constantly mobile, changing audiences. The guided mobility of the viewer through the space includes encounters with an empty amphitheatre, an artificial waterfall that is often dry, a plant nursery and a play area for children with rope walking, slides, etc. The normative disjunctions created between natural resources and construction materials while wishing to engage and entertain audiences of different age groups are representative of the larger complexities of meaning-making and sustainable environmentalism. In addition, the aesthetic and ecological development or presentation of the Tree Park by the Forest Department at a location such as Manipal also requires an engagement with long-standing and ongoing debates about the designation of Ecological Sensitive Areas (ESA) in the Western Ghats[2] (Gadgil et al. 2011: 16–17; Kasturirangan et al. 2023; 11) that has been mired in political controversies. For the purposes of this chapter, such multiple meaning-making and cross-purposive engagements, examined in the context of the Saalumarada Thimmakka Tree Park at Badagubettu in Manipal, have been organized in three sections: narratives of historical and contemporary policies of conservation by the state; narratives of various stakeholders who access the tree park; and narratives gathered through observational fieldwork. By doing so, the study seeks to demonstrate that the Tree Park may be understood as a site where subjects – visitors and inhabitants, human as well as other-than-human – express their liminal positions, and ultimately transform the Tree Park itself into an unruly micro-environment, encouraging a renewed engagement with discourses around green spaces in India.

Greenery as Forest: Narratives of Historical and Contemporary Conservation Policy

Broadly speaking, contemporary environmental policies in India have been shaped by two dominant narratives about wilderness areas and the sustainability of green spaces and trees. The first narrative, which is arguably the dominant one in colonial and post-Independence conservation efforts, views forests essentially as storehouses and reserves of raw material, particularly timber[8] (Agrawal 2005: 21; Margulies 2018: 181–2). Protection of these areas and trees hinges on their being managed sustainably, to ensure a steady supply of materials for the economic development and profit of the State. The second narrative views trees and ecosystems in general are seen as resources of aesthetics and sustenance, whose value depends on their longevity and continued existence. This latter perspective has been a vital rhetoric in shaping the enterprise of eco-tourism, wherein various spaces, their biotic and abiotic components, retain their value as long as they can be sensorially experienced – for instance, as repeatedly attested by photographic images produced and shared by visitors. Both these narratives – of natural resources harnessed for state development or as needing protection and care – can and do co-exist despite appearing to be at cross-purposes if brought to their logical conclusion. Eco-tourism, after all, envisions the greatest profit in resources, such as notable trees or charismatic, individual megafauna, that attract steady clientele to identified locales. On the other hand, the state rhetoric that identifies natural resources for collective 'progress' presents value of natural resources, particularly of trees, as entirely predicated upon their envisioned excision from their environment. Trees, in other words, are valuable not because they are living trees, but because they will one day become timber.

Although manifestations of such contradictory narratives were crystallized into colonial law, and they were inherited by the Indian state post-Independence (Margulies 2018: 183), they have arguably existed since at least the 1000s CE. Alan Johnson, for instance, points out that Sangam Tamil literature views wild spaces as essential to the development and security of the imperial state (Johnson 2023: 5–6); Similar concerns are more bluntly stated in Mauryan writings (Parasher-Sen 2012: 127), such as the *Arthashastra*, a manual of statecraft, which documents, perhaps, an early instance of what Jared D. Margulies calls the 'conservation ideological state apparatus' (Margulies 2018: 182). The authoritarian, imperialistic state emphasized control over territory through utilizing forest resources: both as reserves of timber, and particularly of war elephants, as well as sites of recreation, where the elite would hunt.

The British colonial government, similarly, stood astride these two paradigms: 'managing' forest lands for resources on the one hand, and thus asserting colonial claims over forests in place of local communities (Agrawal 2005: 25), and, on the other, using these forests as recreational spaces for elite hunting expeditions (Rangarajan 2012: 106–7). British authorities, between 1865 and

1927, enacted a series of forest laws whose immediate effect was to control non-elite individuals' access and mobility to wilderness areas. In many cases, colonial policy simply reified and reinforced existing socio-ecological structures; lands that had previously been private hunting reserves of native rulers and landlords came under the ambit – and consequent control – of the Forest Department (Rangarajan 2012: 105). Even lands that are assumed to have been largely under communal control, and perhaps 'inviolate', were not necessarily so: in the upper reaches of the Malabar Western Ghats, for instance, sacred groves were controlled by local landowning elites (Freeman 2012: 174). This policy persisted beyond Independence with minor changes (Margulies 2018: 182), such as the adoption and development of community forestry programmes (Agrawal 2005: 32) and an almost total ban on recreational hunting (Margulies 2018: 182).

In the twenty-first century, the outlook on ecotourism by the Forest Department, and more broadly by the State, remains one inherited from the colonial government. Following Independence, and perhaps more importantly, the launching of Project Tiger in the 1970s, protected green spaces became national parks and wildlife sanctuaries – these quickly became centres of cycles of conflict. Communities living within or on the edge of the park – almost always tribal and lower-caste communities – were evicted, often violently, to make way for a landscape narrated as being pristine and solely the preserve of wildlife (Margulies 2018: 182). Legal access to forests became increasingly restricted to authorities, functionaries of the state and (increasingly) guests who had the material resources to visit such locations. This restrictive policy has, in the past, been met with grassroots resistance movements, particularly in the 1970s, when the Western Ghats, in particular, became the centre of the Appiko Movement. This movement, like its more famous Himalayan equivalent, the Chipko Movement, involved local villagers hugging forest trees in defiance of the Forest Department's policy of cutting them down to supply commercial timber interests (Murthy 1986: 773–4; Mishra 2015: 630).

While the movement was partially successful in preserving forest resources and rights in the 1970s, it has largely ceased to exist since the 1990s, particularly in the Udupi district. Instead, as sites such as the Tree Park demonstrate, existing governmental policy surrounding wilderness spaces has, largely, built upon colonial forest laws and policies.

This has led, in no small part, to a somewhat paradoxical situation that Paul Greenough calls the 'bio-irony' (Greenough 2012: 316) of Indian forests. Greenough points out that official narratives about wilderness spaces in India – particularly well-known national parks and wildlife sanctuaries – are envisioned as being sharply delineated from human communities and anthropogenic activities. Indeed, in most cases, they are defined by a distinct lack of human habitation within their borders. However, they must be accessible to eco-tourists, who come to these locations in order to 'connect with nature'. Catering to eco-tourists allows national parks to survive because it enables them to have cultural currency, generate income for the state, and thus justify their protection. In addition, Greenough points out that this official narrative of wilderness as a space without humans has unexpected consequences. National parks and wildlife sanctuaries may also be used by anti-State actors such as Bodo rebels and Maoist insurgents (Greenough 2012: 324–5). These humans, along with the other-than-human beings which 'stray' out of wilderness areas into anthropogenic environments, may arguably be understood as unruly subjects amenable to disrupting State narratives of 'protecting' green spaces from human 'encroachment'. Such defiances notwithstanding, in terms of State discourse, the survival, protection and narrative currency of many wilderness areas in India depend on their being 'flooded' by relatively privileged and wealthy tourists, on the one hand, and the absence of local communities on the other (Greenough 2012: 331–3). This paradigm, perhaps unsurprisingly, has led to growing cycles of conflict between the State, local communities and wildlife itself.

Similar to these larger green spaces, the Saalumarada Thimmakka Tree Park is arguably a site of both bio-irony and unruliness. The official narrative of the Tree Park envisions it as a site that allows humans to connect with nature, but it constitutes a small enclave of an anthropogenic environment sharply delineated from the green space around it. While advocating the 'services' rendered by natural elements to human communities, its own cultural currency derives from the fact that human visitors to the park use it in unexpected ways, becoming, as shall be shown, unruly subjects in an unruly environment. A park – whether used to refer to a large National Park or to a more contained community area – implies that green spaces have been augmented with infrastructure to facilitate human activities in that environment (Cronon 1995: 12; Spirn 1998: 48). This may range from simply creating walking trails, as with the Tree Park at Manipal, and a significant amount of other built spaces. Visitors and inhabitants at the Tree Park use this infrastructure in ways that are divergent from the stated aims of the Park. Species inhabiting the micro-environment remain embedded in the landscape in spite of having no representation in the official narrative of the State; and concerns of infotainment to visitors, for example, become secondary to leisure disconnected from environmental education.

Names, Causes, Policies and Ecological Entanglements

Thimmakka, a centenarian whose name features on all the Tree Parks set up by the Forest Department, is a crucial conduit and representative of environmental activities and policies in the state of Karnataka. Thimmakka and her husband, Chikkaiah, worked as labourers in the district of Tumkur in southern Karnataka, and although she cannot accurately remember the year of her birth, it is generally estimated that Thimmakka was born in 1911 in the then-Princely State of Mysore. In media interviews, Thimmakka has talked about not being able to conceive in the early years of her marriage and speaks of planting and tending to Ficus trees as though they were her children ('107-Year-Old Environmentalist Planted Hundreds of Trees', 2024). It is from this action that Thimmakka received her moniker of 'Saalumarada': (She) of a thousand trees.

Thimmakka's ecological and affective claim about trees, particularly as 'my trees', is fundamentally different from the dominant State narratives about trees and wilderness areas. From the viewpoint of both ecotourism and sustainable management, trees are essentially viewed as a resource that must be managed for maximum benefit, as decided by the State. In the former paradigm, trees are seen as tools for pleasure and recreation; in the latter, they are viewed as potential bearers of medicine, food and, most importantly, wood; more recently, their value as carbon sinks is also acknowledged. By contrast, Thimmakka's trees, to her, merit intrinsic value as living beings and as her 'children'. In interviews and in her Kannada-language biography, *Saalumarada Sardarini* ('Leader of a Thousand Trees'), Thimmakka extols the value of trees in providing oxygen and supporting biodiversity, particularly *hakki-pakshigalu* (birds), but these 'uses' are prefaced by the assertion that their existence, in itself, is both valuable and unquestionable (Thimmakka, pers. comm. 2018). Her influence is powerful enough to sway public opinion: in 2019, she protested against a State-sponsored road-widening plan that envisioned the cutting down of her trees as a casualty of the project. Thimmakka appealed publicly to H.D. Kumaraswamy, the then-chief minister of Karnataka, and the plan was subsequently altered so that the trees remained unharmed (Shankar 2019). In the same year, she was awarded the Padma Shri ('Awardees' 2019), the fourth-highest civilian honour in India.

Foregrounded against these narratives of resistance to State development activities while simultaneously being feted by the state, Thimmakka emerges as a complex figure. Thimmakka's standing as an environmentalist is based on her articulations around personal aspirations, familial belonging, agential initiatives and collective ecological consciousness. These, in turn, have been mapped upon the government's populist measures and the Forest Department's protection schema, which led to the setting up of Tree Parks, where her name is enshrined and kept continuous with the Saal trees. Saalumarada Thimmakka, however, makes for an uneasy iconography that needs to be interrogated further. Thimmakka's environmentalism can be seen in continuum with a desire to nurture beings; she views her trees, after all, as her children. On the other hand, the environmentalism of the state is self-conscious; the intent is to sustainably manage trees, not necessarily to nurture or tend them, much less view them as children. Similarly, the State views the majority of trees as being held in trust – at least nominally – for the public, and consequently reserves the right to control access. Thimmakka, by contrast, expresses a sense of ownership over the trees she planted, albeit on a stretch of the public road, and does not (or, perhaps, cannot) control access to them. However, at least on a surface level, the connections are obvious: whatever their motivations and their modus operandi, both Thimmakka and the Forest Department have, at least nominally, common concerns – the foregrounding of trees, and more broadly of environmental concerns, to the public. Thimmakka – in lending her name to the park and the scheme – becomes a hypervisible metonym, an unruly and feral subject, not unlike the human visitors to the park, or the street dogs and the invasive plants, that tend to blur or rearrange boundaries to discursive categories such as 'environmentalism', 'conservation' and even 'wilderness'.

In this context, one may argue that unruliness in an environment becomes manifest when the interactions and modes of being of its constituents – whether living beings, abiotic elements or a microregion as a whole – resist narratives that sharply define their place in relation to other beings. They give rise to situations which, as Taylor argues, 'do not have clearly defined boundaries, coherent internal dynamics, or simply mediated relations with their external context' (Taylor 2005: xviii). Unruly elements in an environment thus defy being disciplined – both in their literal bodies and in the ways in which their presence and actions are understood. Instead, they complicate these relationships with other ecological entities and larger-than-individual structures such as the State, forcing them to be dynamic, changing constantly to make sense of ground realities. They are thus 'unruly': their indiscipline, as it were, is not envisioned by others, and more often by the State, which must adapt and become similarly fluid in its dealings with such beings.

Such unruliness implicitly carries within it the possibility of redefining engagements with or between individual species, as well as entire ecological paradigms. In Radhika Govindrajan's study of man-eating leopards in Uttarakhand, for example, the leopards in question are characterized as unruly because they disrupt existing narratives in which leopards and humans are imagined to exist within sharply defined and differentiated environments: the former within wilderness, and the latter within anthropogenic environments (Govindrajan 2015: 33). These leopards are 'ungovernable' (Govindrajan 2015: 35–6); they cross boundaries between wilderness and village landscapes in a literal sense, but also disrupt the discourse that places them outside everyday human lives. It is notable, however, that Govindrajan's characterization of individual species (such as the leopard) as unruly also extends to microhabitats – the latter are seen as unruly, in the sense that they transgress human expectations of these environments and their attendant ramifications.

Similarly, Sunil S. Amrith's book, *Unruly Waters: How Rains, Rivers, Coasts and Seas Have Shaped India's History*, shows that environments may, particularly under anthropogenic pressure, act in unpredictable ways. Human communities and States are forced to re-examine and redefine their relationships with their environments when, under anthropogenic pressures, geopolitical tensions and ecological change, those environments may not only withhold envisioned benefits but, in fact, disrupt and damage human communities (Amrith 2018: 19–21).

While Amrith's study focuses on aquatic tensions, the unruliness that characterizes those waterscapes can be traced to a terrestrial, constructed area such as the Tree Park. Humans as well as other-than-human beings in the Park manifest their unruliness by interacting with the environment in ways not envisioned by the State in constructing and maintaining the Park. More importantly, however, the Tree Park itself becomes an unruly micro-environment. By partaking in these complex interactions between unruly beings, it becomes a site that, in itself, acts in ways not envisioned by the State, with unexpected consequences not only for the ways in which the Park is used but also for the narrative that surrounds and seeks to define the Tree Park itself. In order to fully examine the ways in which the Tree Park in Manipal, along with its visitors and inhabitants, may be understood as defiant of fixed categories, it is useful to understand the narratives that have shaped an understanding of trees and green spaces in India.

Families, Couples and Management Students: Interrogating Narratives of 'Use' in Place-based Experiences

At first glance, the Tree Park in Manipal may seem to partake of a fragmented and nebulous identity. Various sections of it invoke discrete and perhaps disparate histories and narrative identities for the gaze and edification of a visitor. Solitary visitors to the Tree Park are a rarity, and the presence of distinct groups – students, families and couples – suggests the consolidation of community identities. In bringing together these discrete narratives and examining their relationships to these groups, one may interrogate the ideas they invoke and move towards a nuanced reading of the Tree Park as a whole. However, in order to establish a rationale and fully appreciate a scrutiny of select demographic profiles of the park's human visitors, the idea of utility, which is a constant underlying imperative in the presentation of the landscape, needs closer examination.

The site of the Tree Park of Manipal lies on the upward slope of a small hillock, beside the 24-hour-manned gate of the T.A. Pai Management Institute (TAPMI). Other than this imposing and expansive campus, however, it is surrounded by large tracts of open land, some of which fall under the jurisdiction of the Badagubettu village panchayat, and mostly consist of private farmland. Directly behind and beside the park are rolling hillsides covered in wild growth. A species of *Pennisetum*[3] grass predominates, along with several trees and shrubs. During the monsoon, the *Pennisetum* can grow to a height of almost six feet, forming a dense green undergrowth. As the rains recede, the grasses bloom in the winter months and then slowly wither to a yellowish-brown hue.

In contrast to this, the Tree Park itself is marked by its orderly landscape. Winding paths lead from the gate to a bamboo garden at the far end of the park, bypassing small man-made ponds, an artificially fed (and typically non-functional) waterfall, a mini-amphitheatre and several installations depicting arthropod life as well as the cultural life of the Karnataka Western Ghats and coastal region. Dotted throughout the park are saplings of trees deemed culturally or ecologically significant, such as Ficus and sandalwood; small gazebos with benches; and, close to the centre, a herbal garden with several medicinal plants, marked by a board but also, and perhaps more significantly, by a large installation showing a *rishi* pounding medicines, presumably with a mortar and pestle.

The statue of the *rishi*, arguably, deconstructs the straightforward narrative of a sharp delineation between human and wild spaces. At first glance, the statue blandly invokes the broad narrative: it is through anthropogenic activities – and in this case, 'ancient' human knowledge systems – that medicinal plants may be exploited. However, the figure of the ancient Indian hermit intrinsically invokes narratives that question the value and currency of the human landscape. The figure of the hermit is supposed to be one who voluntarily chooses to leave the mores of human society, living in the archetypal *Aranya,* the forest, rather than the *Grama* (Thapar 2006: 24), the village of settled, agro-pastoral human society, either alone or as part of a semi-monastic community of similar individuals (Doniger 2009: 42). In retellings of Indic epics, such as the *Mahabharata* and the *Ramayana,* the *Grama* and *Aranya* are seen, as Romila Thapar notes, as essentially dichotomous; the one is defined by the absence of the other (Thapar 2006: 26). The *Grama* and *Aranya* binary implies an essential difference: the former is defined by human activity and the latter by the lack of human settlement. The hermit, then, becomes an unruly figure who interrogates and redefines boundaries between human and other-than-human communities and structures, in the very act of existing. They are, in other words, feral subjects. We use feral here as defined by Anne Milne: beings whose most salient characteristic is their 'motility', both literal and metaphorical, crossing boundaries between spaces and ideas (Milne 2012: 329). The feral subject's ability to disrupt is also, ultimately, its power to connect; in existing at the interstices of places, ideas and narratives, it emphasizes the ways in which these paradigms may be seen as integrated parts of a whole (Milne 2012: 331). In the environmental context of the Western Ghats, this is perhaps particularly apposite, given that the hills, in an ecological sense, connect the montane world of the peninsula to the marine world of the Arabian Sea; the coastline of Karnataka alone has ten major estuaries, including two (the Swarna-Seetha and Kundapura estuaries) that fall into Udupi district, less than 50 kilometres from the Tree Park in Badagubettu. The feral animal's invocation of connection – across micro-environments, human and wild spaces, and ways of understanding space – allows for the linkage of multiple complex narratives about these spaces.

In many ways, the layout and landscaping of the park similarly represent the multiplicity of narratives about forests and trees that exist. The bamboo garden, with its many varieties of bamboo in neat enclosures, each with a board that mentions not only their English and Latin names but also pronounces the benefits commonly derived from each bamboo by human communities. Several species are not native to the Western Ghats, with origins as far afield as northeastern India and Java. Together with the information boards that talk about the 'services' rendered by green spaces (alongside a board extolling services rendered by marine ecosystems), the bamboo garden represents, as it were, the ur-narrative of the Forest Department: it suggests that forests and trees are meant to be utilitarian resources, managed and protected insofar as to ensure that they can provide a constant and renewable source of raw materials. This much older narrative then attaches added concerns of the twentieth and twenty-first centuries: maintaining ecological balance, acting as carbon sinks and fending off the spectre of climate change. The juxtaposition of the bamboo and herbal gardens represents an uneasy continuum of divergent narratives. If the bamboo garden represents the ur-narrative of the State, the herbal garden invokes an epistemic paradigm in which colonial narratives of wilderness are being challenged and rewritten.

The idea that forests contain medicinal plants is not new; in recent times, however, it has gained currency and controversy, following the corporatized overexploitation of medicinal plants in wild spaces (Friere et al. 2023: 209–12). Essentially, however, the medicinal garden invokes the classical colonial narrative, of forests being essentially resources that must be sustainably exploited. Against this background, then, the Tree Park emerges not as an ecological isolate, but as a link that exemplifies the ways in which ecological and eco-touristic categories – trees, forests, park, hills – blur their environmental and social boundaries.

These quilted narratives of 'beneficial' medicinal plants or foliage, with features that appeal to commonly held cultural notions of beauty and aesthetics, are ultimately uni-directional in their acute awareness of the human audience or eco-tourist. In contradistinction, human visitors to the park do not fit easily into the Forest Department's broad category of 'urban dwellers', and indeed few, if any, make use of the Tree Park's other envisioned utilities, such as the jogging track and the amphitheatre. A fair number of visitors include extended and nuclear families – which include a noticeable number of Muslim families clad in burqas or niqabs that readily mark them as such – and young heterosexual couples, primarily from the region, where social restrictions and a lack of private spaces likely make the park an attractive destination. Another significant demographic in the park is students from TAPMI, the adjacent management school, many of whom bypass actually visiting the park and let themselves directly into the canteen on its periphery.

In order to fully understand these interactions, the authors visited the Tree Park over a two-week period, for up to three hours at a stretch. They visited on all days of the week except Monday, when the Tree Park is closed. The observational ethnography was conducted in the month of September, following the monsoon, when the flora of both the Tree Park and the surrounding terrain had dense foliage, and several species of flowering plants, such as *Ipomoea*, Hibiscus and Morning Glory, had begun to flower. While human-to-human interactions were limited, and formal interviews were not conducted, the authors relied on observational ethnography to document visitors, inhabitants, workers and the interactions amongst them at the park. Such an approach suggested itself because the intent was to observe the uncontrolled, day-to-day mechanisms through which beings at the Tree Park built their narratives of place by interacting with the environment and each other.

From the resulting ethnographic notes, the authors were able to broadly profile the majority of visitors. Though this approach carries the risk of generalizing visitors and their interactions, it is useful because it takes cognizance of the beings who exist in that terrain over the majority of that time. By relying on easily observable markers of identity – uniforms, traditional clothing and language-use – it emerged that the majority of visitors to the Tree Park could be broadly classified into three main groups: Muslim families from

Udupi and its locales, students from the nearby management institute and young heterosexual couples.

At first glance, these three demographics could not seem more unlike each other. Yet, unsurprisingly, and in different ways, the Tree Park offers them a space that is, arguably, safe precisely because it straddles the boundary between an anthropogenic and a wilderness environment.

Udupi, with its reputation as a temple town, is primarily known for its religious locales and rituals, with beaches, malls or cinema halls becoming more secular forms of entertainment. In addition, the entire coastal Karnataka region – the districts of Dakshina Kannada, Udupi and Uttara Kannada – has, in recent decades, been a political stronghold of a national party with Hindu right-wing ideology. The urbanscape of Udupi features a multitude of celebratory markers (banners, flags, icons, devotional songs on speakers) of the Hindu religion. Such obvious polarization and religious colouring of public spaces and activities pose challenges to creating comfortable, accessible spaces for group outings for minority communities. At the Tree Park, by contrast, Muslim families may be understood as having access to a different pace of leisure within a secular idiom. Media reports, as well as scholarship, underscore the fact that Muslims in the Udupi district constitute a large population who are often explicitly or implicitly targeted, as a community, by Hindu right-wing ideologues, private individuals and groups. The Muslim community in and around Udupi, thus, is often overtly made aware of their identity as a minority.[4] Against this socio-political background, the Tree Park represents a space where one's identity as a Muslim is, arguably, subsumed to other identities and is consequently, in some ways, a safe space to exist as a visibly Muslim person or family. Often, in these outings, adult men are absent, or, if present, outnumbered by women and children. The women tend to relax in the shade of the gazebos, while the children play on the lawn or, accompanied by one or two adults, explore the park.

The statue of the *rishi* notwithstanding, the Tree Park offers a space where the visitor is guided to identify in terms of species (human) rather than religious community (Muslim or Hindu). The nature of the other sculptures at the Park is also, perhaps, significant: they are either broadly considered 'cultural' (rather than communal) in nature, such as the sculptures depicting *Pili Vesha* (the tiger dance) and *Kambala* (traditional buffalo racing), or educational, like sculptures of frogs, mushrooms and invertebrates.

For young couples, similarly, the park offers relative privacy and safety. While the nearby town of Manipal is relatively liberal in comparison to the much more socially conservative and traditional Udupi, it offers only a handful of 'safe' places (other than private homes) where a couple might find both a site away from disapproving eyes and a place where they need to spend very little money in order to gain entry and stay as long as they would like. Neither families nor couples at the Tree Park make any effort to conceal themselves and routinely spend almost three hours at the park, often moving between gazebos or taking long walks through it.

By contrast, the students of the nearby management institute, who let themselves into the canteen, are often more brief visitors: as long as it takes to smoke a quick cigarette and drink a cup of tea or coffee along with it. Sited close to the entrance, the canteen is accessible both from within the park and from outside it, and – in common with many similar establishments in Manipal and Udupi – features a small, hidden and quasi-legal smoking area. In this, the canteen seems to fulfil an almost archetypal role: poised on the literal edge, between presumed civilization (the TAPMI campus) and claimed wilderness (the Tree Park and the panchayat lands beyond), it offers the comforts of a human environment: ready food, service staff and consumables; as well as the freedom suggested by a wilderness, a place separated from the strictures of an anthropogenic environment, where control – in this case, by college authorities – is interrogated and circumvented.

In many ways, this is perhaps a classic eco-tourism experience: fundamentally, 'eco-tourism' suggests a focused engagement with the natural and the other-than-human, but not necessarily devoid of human comforts, to a lesser or greater degree. The many human visitors to the park thus become unruly participants. They act in ways not wholly envisioned by the Forest Department when setting up the park, though, in many cases, their goals align with those of the Tree Park. While they may not be actively educating themselves on varieties of bamboo or medicinal plants, for example, it may be argued that they are nonetheless connecting with nature. In viewing and using the Tree Park as a place of leisure, they are not only interrogating how such 'feral' places – neither entirely anthropogenic nor entirely wild – are understood and used, but also building affective attachments to the Tree Park.

In other words, they are engaging in what the ecocritic Lawrence Buell calls place-making (Buell 2001: 13–15). Buell argues that this process is of vital importance in the larger environmental and ecocritical perspective: we choose to protect spaces which transcend into places; sites to which we develop affective attachments (Buell 2001: 15). The unruliness of the Tree Park's subjects, if anything, perhaps makes their place-making more robust: They interrogate the State's narrative of the ways in which the Park is used, and make it their own, as it were, by redefining the space through their actions and emotional attachments. The visitors and inhabitants of the park establish affective connections to the terrain: as a place of leisure, of secularism, as a space shielded, to some extent, from social and governmental surveillance, as a space where interactions between human and other-than-human beings are foregrounded. While some of these affective connections have explicit or implicit State sanction – of viewing the Tree Park as a place of leisure, for example – many do not. For instance, feral life-forms, which do not feature in the Forest Department's narrative of the Tree Park, nonetheless constitute an important element of visitors' and inhabitants' affective connections. Consequently, visitors' place-making, in many ways, questions the modes through which the State imagines and narrates the Tree Park itself.

Feral, Domestic, Migrants and Such Life Forms among Trees

The Tree Park, while signalling a curation of various species of trees and foliage, also provides a habitat to a wide spectrum of residential as well as migratory flora and fauna – feral in its biological sense. For instance, several winged species, which fly in and over boundaries between panchayat land, open fields, the Tree Park and the TAPMI campus, for instance, embody the 'motility' (Milne 2012: 330) described by Anne Milne, perhaps better than many other beings. Milne argues that the defining characteristic of a feral being is its ability to cross otherwise strictly delineated boundaries, whether physical (between wilderness and human habitation), scientific (wild, domestic, tame) or ideological (both cohabiting with humans while not being dependent on them in order to survive). However, this section focuses on two unruly subjects of the Tree Park: the feral dogs who live within and around it, and the invasive plants, such as Morning Glory creepers and some species of Touch-Me-Not, which spring up within the park, particularly during and after the monsoon. These two entities – dogs and plants – are of significance because, unlike birds or even more 'dangerous' animals such as snakes, they persist in inhabiting the locale in spite of being absent in the State narrative of the Park. Animals such as snakes and invertebrates in the Western Ghats frequently encounter humans, live in shared environments and sometimes enter human homes – these are championed by the Tree Park in the signages. Large display boards explain how important snakes are to the ecosystem, as also invertebrates and frogs, in addition to large concrete sculptures. One of the latter depicts a life-like python sculpture, which, from its position within a grotto of rocks just above a small wooden bridge, has notably startled several visitors.

By contrast, the dogs and invasive plants are both unmentioned and ubiquitous. In biological terms, a feral individual belongs to a species that has undergone genetic change due to human intervention – generally through artificial selection – but which has 'reverted' to a wild state; it is, in other words, domestic but not tame, or, in the case of plants, it is a cultivar but not cultivated (Pollan 2002: 5). The modes of being of the Tree Park's invasive plants and dogs also discomfit even such broad categories as 'feral'.

The dogs at the Tree Park – in common with many 'street dogs' across the Indian subcontinent, who are fairly domesticated but live on the streets – tend to be a focus of both human affection and fear. Any perceived feral-ness hinges upon their being owner-less, their living in conditions unattached to any particular human family or individual, rather than upon their acting archetypally wild. The Tree Park dogs are noticeably comfortable around humans and will often seek out human visitors to the park if they have food, or even affection, at hand. The fact that they continue to do so, and to try their luck, as it were, with multiple visitors suggests that this interaction is both ongoing and has its rewards: at least a few human visitors to the Tree Park dispense food and physical affection, building and maintaining a human–animal relationship of biophilia – an affinity to other-than-human beings (Wilson 1986: 11), in the words of E.O. Wilson. At least one puppy wears a ragged blue collar, which suggests that it enjoys some amount of protection from, though not adoption into, a human family.

Individuals such as this animal further complicate the relationships between humans, other-than-human beings and the places they share. Their existence in the park and their interactions with other beings are unaccounted for in the official narrative of the Tree Park, in which dogs are not even considered possible denizens of the site. However, they are clearly of significance: in their interactions with at least a few human visitors, they act as a vital link in building affective attachments to the park. The dogs, being territorial, remain attached to the Tree Park itself as a part of their presumed home range; and for people who interact with them – at least those who interact positively with them – they act as living, motile and memorable markers of Place.

Like the dogs, the invasive plants that grow in the Tree Park, too, straddle lines between 'feral', 'cultivated' and 'cultivar'. In the monsoons, the entire site – both within and outside the boundaries of the Tree Park – sees the almost exponential growth of various species of Morning Glory (*Ipomoea* species). Largely originating from Central and South America – where the climate echoes that of the Western Ghats – these flowering creepers form dense, fast-growing thickets and can crowd out native plants and saplings, cutting off light and space.

Discussions around invasive species often posit them literally as invaders: unwelcome aliens with an intent to conquer. The language used to describe them – at times violent and xenophobic – can often seem as if it has been displaced from similar discussions about human populations of immigrants and minorities (Mitchell 2024: 76). The fact remains, however, that most invasives have been introduced purposefully rather than accidentally to landscapes through human intervention. Lantana, for instance, in many ways the most visibly problematic and well-known of India's invasive plants, was introduced as a garden plant. There is documentation of it being introduced separately, at opposite ends of the country on two separate occasions: in Coorg (modern Kodagu district in Karnataka) in the 1860s, and in 1911 in Calcutta (modern Kolkata). Thereafter, it spread through the landscape, 'indiscriminately taking over farm and pasture, fallow land and forest' (Hiremath et al. 2005: 33)

A similarly documented micro-history of *Ipomoea* does not exist, but it is not unreasonable to assume that it followed a similar trajectory: as 'escapees' from cultivated gardens. This is lent credence by the fact that, with their bright flowers and ease of cultivation, Morning Glories can, and do, often revert to being cultivated in gardens. Such is certainly the case in the Tree Park, where the monsoon growth of creepers is likely tacitly encouraged. A pair of Morning Glories that had climbed up a gazebo pillar, for instance, had been left to grow, perhaps because their flowers had bloomed at an aesthetically pleasing conjunction with the gazebo itself. In other parts of the park and around its enclosing walls, these creepers and other invasives grow in dense abundance.

Like the dogs, these creepers defy easy categorization in their role as inhabitants or occasional migrants into shared space. They may gain a root-hold in the Forest Department's narratives as plants, but they sit uncomfortably in that paradigm on account of being 'displaced species' (Mitchell 2024: 104), in an environment not native to them and upon which they have a deleterious effect. They are 'wild' in the sense of having been germinated by natural, rather than anthropogenic, causes, but they re-enter the world of human environments by being encouraged to grow by human communities, protected and even embedded into the visual narrative of the Tree Park, as in the case of the two creepers at the gazebo.

It is, arguably, difficult to observe direct, focused interaction between human visitors to the Tree Park and invasive plants there. Intentional interactions with vegetal life are dominated by the more valorized species that have been intentionally planted there, such as the bamboos, the ficus, and some fruit trees. However, it is, perhaps, not unreasonable to assume that, in forming such a visible part of the landscape, the various invasives play a significant role in the way the Tree Park and its surroundings are imagined and narrated into human lives. The unruliness of the *Ipomoea* creepers lies in their unexpected embedding into the environment of the Tree Park, implicitly challenging the State narrative of curating trees to facilitate an orderly and disciplined interaction between human and vegetal beings. The idealized Tree Park, in the eyes of the State, is one in which, arguably, all vegetal beings exist in a controlled, orderly manner: they are planted and tended to by the Forest Department, and their interactions with visitors follow this. They exist in order to explicate a narrative in which other-than-human beings exist within sharply defined spaces – Reserved Forests, sanctuaries or Tree Parks – and express the ability of the Department to exert control over these areas. However, because the *Ipomoea* is not anthropogenically planted, remains untended by the Forest Department, and persists in the terrain of the Tree Park while forming a significant element of visitors' interactions with plants, it comes to represent a distinctly unruly vegetal element in that narrative.

Concluding between, alongside, and within Parks and Boundaries

In environmental humanities scholarship over the past two decades, much attention has been devoted to large-scale, and indeed global, solutions to environmental problems (Shiva 2005: 23). Increasingly, however, recent scholarship has moved towards reframing environmental and ecological problems – and consequently their study and possible solutions – to focus on micro-regions and micro-histories of place (Shahabuddin et al. 2017: 12; Govindrajan 2018: 6), allowing these phenomena to be understood in terms of the communities, living beings and geographies that they most directly impact. In addition to the measurable quantums through which environmental change may be measured – such as economic, climatological and biological data – studies that take into account, and centralize, affective relationships in such geographies aid in making sense of narratives of place. These – in spite of their particularity – may be extrapolated to bear upon other sites of environmental importance. This chapter, then, by presenting an affective, place-centred understanding of a singular site, draws together these strands of scholarship on affective responses, observational ethnography and micro-regions, demonstrates the ways in which narratives of place may be reframed in order to make sense of socially and ecologically complex and significant environments.

This study examined multiple narratives attached to and emanating deliberately or otherwise from the Saalumarada Thimmakka Tree Park in Badagubettu, Manipal, Karnataka. Established and maintained by the Karnataka Forest Department, this small, semi-natural space, with curated species of trees, features infrastructure (like gazebos, a small waterfall, a canteen and an amphitheatre) and participates in environmental education while promoting ecotourism. This is achieved through propagating the importance of natural resources and the conservation as well as enjoyment of the environment in general, and of trees in particular. This chapter posits that the Tree Park may be read as a discursive space in which State narratives about human relationships with the environment are interrogated and redefined. The inhabitants and visitors of the Tree Park, it is suggested, are 'unruly' in the sense that their modes of existence within the park are not envisioned by official discourse, and in so doing, they establish affective attachments to the site, imbuing it with eco-cultural value.

Historically, the State's impulse has been to enforce the enclosure of green spaces and restrict human access to them, because trees are viewed fundamentally as resources to be sustainably exploited and held in trust for the public. In this narrative, humans and other-than-human beings are seen as inhabiting sharply delineated habitats, the latter pointedly anthropogenic, the former supposedly 'natural'. The Saalumarada Thimmakka Tree Park Scheme, however, hints at a growing change in this official discourse. Named after an unlettered woman from a rural background whose nurture of trees despite limited material resources remains legendary, it represents a distinct approach in governmental policies around green spaces adjacent to urban areas. It identifies and fences a space for the visiting public to interact with nature at a nominal price – albeit a version of nature that is curated by the State – and encourages a 'connect' through multi-usage: a space to interact with and be educated about the natural world, accompanied by opportunities for exercise, leisure and even performances. In doing so, the first-hand experiences and narratives of visitors become entwined with more official narratives of environmentalism, tree conservation and 'connecting with nature'.

These different functions of the Tree Park turn it into a site where many beings express their unruliness; their interactions with the environment and with each other are not envisioned by the official discourse surrounding the park. Similarly, the sculptural installations depicting human activities with other animals, such as the buffalo race in paddy fields (*Kambala*) or cultural performances where humans take on the garb of other animals, such as jungle cats (*Pili Vesha*), hint at the ways in which the Tree Park does not fall easily into simple paradigms of 'natural' or 'anthropogenic'. Thus, for Muslim families and young couples from Udupi, as well as management students from TAPMI, the Tree Park is a space where restrictive State and social narratives may be interrogated and defied. The feral dogs and invasive plants of the site, likewise, are embedded into the environment, interacting with the visitors and installations; their modes of being in the park have no place in official discourse and, by their existence in the landscape, they complicate the ways in which the park may be read and understood.

The currency and significance of the Tree Park, in the broader context of environmentalism, however, lie in this unruliness of its participants. The stakeholders of the park – the human visitors, the Forest Officials and workers who govern it, and the other-than-human-beings that inhabit it – develop affective attachments to the site because it adapts to, and implicitly encourages, the building of an affective relationship, the unruliness of this process notwithstanding. As they interact with the space and the beings within it, they contribute to its developing 'value' as a place that encourages human interactions with other life forms within a given landscape. This, in turn, adds significance to notions of leisure in a semi-natural environment in the Western Ghats, and to viewing trees and wildlife as having intrinsic value by virtue of their significance to human modes of existence – not only as entities that render 'services,' as the State narrative asserts – but because they become involved in complex relationships between species, environments and narratives of place.

Notes

1 The entry fee at the Tree Park is small: twenty rupees per head for adults and five rupees for children. Connectivity to the park, arguably, exercises a greater influence on the socio-economic profile of visitors to the park.

2 Both of these reports were commissioned by the Government of India, and constituted the first environmental survey of the entire Western Ghats region, across six states. Although the two reports' recommendations differed widely in some respects, both recommended that large sections of the Western Ghats be designated Ecologically Sensitive Areas, which would preclude certain industrial activities in those areas.

3 *Pennisetum* grasses are commonly known as fountain grasses. They tend to form thick clumps, arising from closely growing roots, and usually have an underground bulb, small hairs along the stems, and long, bulrush-like flowers.

4 For a generalized understanding of Muslim positions in the Indian urban space, see Laurent, Gayer, and Christophe Jaffrelot (eds). *Muslims in Indian Cities: Trajectories of Marginalisation*. London: C. Hurst and Co., 2012. For more specific information about Muslims in Udupi and Dakshina Kannada districts, see Vaddiraju, Anil Kumar. *'Karnataka and Andhra Pradesh: Urban Primacy and Urban Centralisation' in Urban Governance and Local Democracy in South India*. New Delhi: Routledge, 2021. For media reports, see 'Udupi Police File Case against Social Media Account for Derogatory Post against Muslims' in *The Hindu*, 15 July 2024. and 'Udupi Muslims Urge Karnataka CM to Withdraw Ban on Hijab in Classrooms' *The Hindu*, 26 December 2023. Also 'Udupi Temple Bans Non-Hindus from Setting Up Stalls during Annual Festival' *The News Minute*, 23 March 2022.

References

'107-Year-Old Environmentalist Planted Hundreds of Trees'. *YouTube*, NDTV, youtu.be/LXptbYaWag0?si=XR5SQJoYbuwglyMJ (accessed on 4 November 2024).

Agrawal, Arun. 2005. *Environmentality: Technologies of Government and the Making of Subjects*. Durham: Duke University Press.

Amrith, Sunil S. 2018. *Unruly Waters: How Mountain Rivers and Monsoons Have Shaped South Asia's History*. New York: Basic Books.

Aranya.Gov.In. 'Tree Park Scheme'. Karnataka Forest Department, aranya.gov.in/aranyacms/(S(mg0j4gm4slw155gmmc4mxeiv))/English/IndividualScheme.aspx?TpfCI446hYpiHVjq11kX6w=%3D (accessed on 4 November 2024).

'Awardees'. *Padma Awards*, Ministry of Home Affairs, Government of India. www.padmaawards.gov.in/padmaawards/padmaawardees (accessed on 4 November 2024).

Buell, Lawrence. 2001. *Writing for an Endangered World: Literature, Culture, and Environment in the US and Beyond*. Cambridge, MA, and London: Harvard University Press.

Centre, U. W. H. n.d. 'Western Ghats'. Maps – UNESCO World Heritage Centre. https://whc.unesco.org/en/list/1342/multiple%3D1%26unique_number%3D1921.

Cronon, William. 1995. *Uncommon Ground: Rethinking the Human Place in Nature*. New York: W.W. Norton.

Doniger, Wendy. 2009. *The Hindus: An Alternative History*. New York: Viking Press.

Dubash, Navroz K. 2007. 'The Local Politics of Groundwater in North Gujarat'. In *Waterscapes: The Cultural Politics of a Natural Resource*, edited by Amita Baviskar, 40–65. Ranikhet: Permanent Black.

Freeman, Richard. 2012. 'Folk Models of the Forest Environment in Highland Malabar'. In *India's Environmental History: Colonialism, Modernity, and the Nation*. 741–56, edited by Mahesh Rangarajan and K. Sivaramakrishnan, 169–97. Ranikhet: Permanent Black.

Freire, Ague, Fernanda Moura Fonseca Lucas, Francival Cardoso Felix, Kyvia Pontes Teixeira das Chagas and Allana Katiussya Silva Pereira. 2023. 'Plants That Heal: The Sustainable Exploitation of Medicinal Resources in Brazilian Forests'. In *Forest Science: Environmental*

Footprints and Eco-design of Products and Processes, edited by E. C. de Souza and S. S. Muthu, 207–45. Singapore: Springer.

Gadgil, Madhav. and Western Ghats Ecology Expert Panel. 2011. *Report of the Western Ghats Ecology Expert Panel*. New Delhi: Ministry of Environment and Forests, Government of India.

Govindrajan, Radhika. 2015. 'The Man-Eater Sent by God: Unruly Interspecies Intimacies in India's Central Himalayas'. *RCC Perspectives* (3): 33–8.

Govindrajan, Radhika. 2018. *Animal Intimacies: Interspecies Relatedness in India's Central Himalayas*. Chicago, IL: University of Chicago Press.

Greenough, Paul. 2012. 'Bio-Ironies of the Fractured Forest: India's Tiger Reserves'. In *India's Environmental History: Colonialism, Modernity, and the Nation*, 316–57. 888–927. Ranikhet: Permanent Black.

Hiremath, Ankila J. and Bharath Sundaram. June 2005. 'The Fire-Lantana Cycle Hypothesis in Indian Forests'. *Conservation and Society* 3(1): 26–42.

Johnson, Alan. 2023. *India's Forests, Real and Imagined*. London: Bloomsbury Academic.

Kasturirangan, Krishnaswamy. and Committee on Western Ghats Ecology Expert Panel. 2023. *Report of the High Level Working Group on Western Ghats*. Ministry of Environment and Forests, Government of India.

Margulies, Jared D. 2018. 'The Conservation Ideological State Apparatus'. *Environment and Society* 16(2): 181–92.

Milne, Anne. 2012. 'Fully Motile and AWAITING FURTHER INSTRUCTIONS: Thinking the Feral into Bioregionalism'. In *The Bioregional Imagination: Literature, Ecology and Place*, edited by Tom Lynch, Cheryl Glotfelty and Karla Ambruster, 329–47. Athens, GA: University of Georgia Press.

Mishra, Renu. 2015. 'Poltics of Environment and Development'. *The Indian Journal of Political Science* 76(3): 629–31. Stable URL: https://www.jstor.org/stable/26534901.

Mitchell, Audra. 2024. *Revenant Ecologies: Defying the Violence of Extinction and Conservation*. Minneapolis, MN: University of Minnesota Press.

Murthy, K. S. Dakshina. 1986. 'Politics of Environment'. *Economic and Political Weekly* 21(18): 773–5. Stable URL: http://www.jstor.org/stable/4375626.

Parasher-Sen, Aloka. 2012. 'Of Tribes, Hunters, and Barbarians: Forest Dwellers in the Mauryan Period'. In *India's Environmental History: From Ancient Times to the Colonial Period*, edited by Mahesh Rangarajan and K. Sivaramakrishnan, 127–51. Ranikhet: Permanent Black.

Pollan, Michael. 2002. *The Botany of Desire: A Plant's Eye View of the World.* New York: Random House.

Rangarajan, Mahesh. 2012. 'The Raj and the Natural World: The Campaign against "Dangerous Beasts" in Colonial India, 1875–1925'. In *India's Environmental History: Colonialism, Modernity, and the Nation*, edited by Mahesh Rangarajan and K. Sivaramakrishnan, 95–142. Ranikhet: Permanent Black.

'Saalumarada Thimmakka Vruksha Udyana Yojane'. Karnataka Forest Department. https://aranya.gov.in/aranyacms/(S(2bli1w5sdzushslzsri3cbi0))/English/IndividualScheme.aspx?q0t0dYHZU8PhPzPshKCh7A==.

Shahabuddin, Ghazala, Mysore Doreswamy Madhusudan and Mahesh Rangarajan (editors). 2017. *Nature without Borders*. Ranikhet: Orient Blackswan.

Shankar, B. V. Shiva. 2019. 'HD Kumaraswamy Intervenes to Save Trees Planted by Salumarada Thimmakka'. In *The Times of India*. Bengaluru, 3 June.

Shiva, Vandana. 2005. *Earth Democracy: Justice, Sustainability and Peace.* London: Zed Books.

Spirn, Anne Whiston. 1998. *The Language of Landscape*. New Haven: Yale University Press.

Taylor, Peter J. 2005. *Ecology, Interpretation, Engagement*. Chicago, IL: University of Chicago Press.

Thapar, Romila. 2006. 'Forests and Settlements'. In *Environmental Issues in India: A Reader*, edited by Mahesh Rangarajan, 33–41. New Delhi: Pearson Education India.

Wilson, Edward Osborne. 1986. *Biophilia*. Cambridge, MA, and London: Harvard University Press.

Section II

Narratives of Gender

5

(Re)writing 'The Perfect Man': Reading De-idealized Masculinities of Rama and Lakshmana in Samhita Arni's *The Missing Queen* and Pervin Saket's *Urmila*

Aashna Viswanathan and Beniamin Kłaniecki

Introduction

Over the past few decades, the cultural legacies of the *Ramayana* have been mobilized to depict Ram and Lakshman as archetypes of the 'ideal' Indian man in popular culture and politics. This depiction of masculinity may be seen in the discursive and visual stylization of Ram as the *maryada purushottam* (the 'ideal' man) (Richman 2001: 15), and the *rashtra purush* – 'the god-king [refashioned] into the ideal national man' (Brosius 2004: 95) – in contemporary cultural narratives, such as television shows and calendar art. Consequently, this (ideal)ized masculinity has not only become a canonized blueprint for morality and nationalist devotion but also a means to control and maintain hierarchies in class, caste and gender through various social, often intersecting, power relations (Richman 2001: 10–12). In response to this trend, we examine post-millennial feminist retellings of the *Ramayana* that engage from a feminist position with the epic's twentieth- and twenty-first-century narrative legacy. While these retellings are predominantly concerned with feminist themes, they provide particularly valuable insights into the process of constructing masculinity, which, as we observe, remains unproblematized in the currently dominant *Ramayana*-inspired narratives.

In the contemporary context from which these retellings emerge, it is not feasible to speak of a single *Ramayana*, let alone a definitive version of the classical *Ramayana*. We, therefore, propose to read the *Ramayana* in this chapter as an amalgam of various undercurrents and strands of popular retellings of this epic, disseminated through print, television and oral circulation, that resonate with various audiences across modern India. The contemporary feminist retellings that we examine challenge these versions predominantly by engaging with the fictional and lived realities of Indian women. Both Samhita Arni's *The Missing Queen* (2013) and Pervin Saket's *Urmila* (2016) bring the often-overlooked stories of the female characters in the *Ramayana* to the forefront, while employing the epic's political power to offer counter-narratives to the nationalist and other mainstream retellings.

We suggest that, as feminist retellings, both novels not only centre female perspectives but also effectively problematize the construction of Ram's and Lakshman's idealized masculinities, which dominate both postcolonial and nationalist-leaning versions of the *Ramayana*. Specifically, we argue that the novels present the two brothers as embodied men, revealing the disjuncture between the idealized representations of masculinity and the lived experience of men. Reimagining the brothers' traditional roles and their *dharmas* – Ram as king and Lakshman as the king's brother – Arni dissects the political nature of Ram's masculinity, while Saket offers a 'domestic' critique of Lakshman's masculinity as the absent husband.

In our analysis, we understand masculinity as defined by R.W. Connell (1995): 'a place in gender relations, the practices through which men and women engage that place in gender, and the effects of these practices in bodily experience, personality, and culture' (Connell 1995: 71), articulated through relations of power, with respect to both women and other men. In *Masculinities*, Connell (1995) famously distinguishes four kinds of masculinities based on their position within power dynamics among men and the socially upheld notions of male power. By viewing the lived experiences of men as embodiments of specific masculinity models situated along a power hierarchy, Connell adopts an essentialist, descriptive approach to gender. This contrasts with Judith Butler's famed constructivist view, which posits that gender is a performative social construct predating the body: 'gender reality is performative, which means, quite simply, that it is real only to the extent that it is performed' (Butler 1988: 527). This ontological divergence is reflected in how both novels deconstruct masculinity – not as an inherent essence but as a set of pre-existing expectations – and gesture towards what else these masculinities might become through the characters of Ram and Lakshman.

In Connell's model, the concept of hegemonic masculinity has its roots in Antonio Gramsci's theory of hegemony, which originates from Marxist thought. Gramsci defines hegemony as 'the ideological predominance of the cultural norms, values, and ideas of the dominant class over the dominated' (Woolcock 1985: 204). Tied to the Marxist notion of class struggle through its origins, hegemonic masculinity foregrounds the exercise of power through gender stylizations within specific geographical, temporal and political settings (Connell and Messerschmidt 2005: 829–3). As such, hegemonic masculinities are represented by individuals who exhibit attributes considered masculine by culture and society, and maintain proximity to institutions of power. Further from power are complicit, marginal and subordinate masculinities, respectively (Connell 1995:76–81).

However, one key critique of Connell's framework is that hegemonic masculinity is largely unachievable for most men (Peltola 2022: 5). In the case of Ram's and Lakshman's portrayals in postcolonial and subsequently post-millennial India, their gender construction extends beyond the scope of Connell's framework and its critiques; therefore, the brothers' masculinity cannot be described as hegemonic in Connell's sense. Instead, we propose that the masculinity of Ram and Lakshman in the contemporary Indian context should be read as idealized masculinity – a symbolic site that draws on complexities attached to gender identities (Butler 2007: 1993). These identities are invariably mutable across time and place, depending on the socio-political context.

While this idealized masculinity cannot be humanly embodied, it exerts immense power over the collective visual imagination and political agenda, as evidenced by the re-stylizations of Ram at the turn of the millennium in response to community and religious affiliations. Tapping into the mutable nature of the epic's tradition, these narratives tend to produce an ideological order to be reflected in real life, with the aim of regulating and redefining India as a 'pure', monolithic nation. The epic's central preoccupation with 'male identity and female fidelity' (Doniger 2010: 227) has facilitated versions of the epic that mass-produced hypermasculine iconizations of Ram and Lakshman, used to build foundations of 'idealized' identities of heroic masculinity. We accept, therefore, that the masculinities of Ram and Lakshman have been central to the postcolonial period and more recent nationalist discourses, and that they continue to serve as potent carriers of political power.

Gender in the Narrative Traditions of the *Ramayana*

As previously noted, contemporary feminist retellings of the *Ramayana* emerge from a broader network of the epic's earlier versions, including the postcolonial renditions that idealize traditional masculinity and emphasize 'moral and ambiguous deeds' (Richman 2001: 10), alongside Ram's and Lakshman's devotion to their *dharma*. Many of these feminist retellings now challenge patriarchal notions of masculinity, foregrounding gender as a central site of critical engagement. In doing so, they often echo, even if indirectly, earlier narratives more attuned to the plight of their female characters. For instance, Shulman's (1991) analysis of Kampan's Tamil retelling, *Iramavataram*, shows Ram's excursion to save Sita as a way to redeem himself rather than to fulfil his duties as a husband and save his wife from Ravana's confinement; eventually, he blames Sita for abandonment while neglecting Sita's experience of abduction. In their research on the oral traditions of rural women singing the *Ramayana*, Velcheru Narayana Rao (1991), Nabaneeta Dev Sen (2008) and Namita Gokhale and Malashri Lal (2009) point to increased engagement with female experiences such as Kausalya's pregnancy, Sita's puberty and wedding, Surpanakha's revenge and Urmila's sleep, thus decentralizing the adventures of Ram and Lakshman that bind the *Ramayana*.

While there is a heterogeneous narrative tradition of retelling the *Ramayana* across languages and time (Richman 1991), the last few decades, according to Anindita Banerjee (2011), have witnessed a 'hegemonic, exclusivist narrative of the *Ramayana* that has played a fundamental role [...] in both domestic and public spheres, protecting and purifying an explicitly feminized familial and national body from a religious/ethnic other, and establishing the utopian order of *Ram-rajya* (Ram's kingdom)' (2011: 125). These dominant narratives in the post-2000 literary scene tend to celebrate heroic masculinity while marginalizing other masculinities, reflecting the coexistence of the perceived resurgence of Hindu affiliations on the one hand and emerging subcultures, on the other hand, providing spaces for assertion and exploration of marginalized genders and sexualities.

The revival of rewriting the epic in the popular literary sphere of post-2000 India has primarily comprised narratives that foreground the suppressed voices of women, including the writings of Chitra Banerjee Divakaruni, Kavita Kane, Anuja Chandramouli, Devdutt Pattanaik, Smriti Dewan, Manini J. Anandani, Amish Tripathi, Malika Sengupta, Amit Majumdar, Anand Neelakantan, Samhita Arni, Pervin Saket and Koral Dasgupta. Their renderings have challenged representations of female characters prevailing in the dominant versions of the *Ramayana*, where the women are represented either as the *pativrata*, the 'ideal' or virtuous wife (embodied by Sita) or as the villainess or monstress (such as Surpanakha). These feminist retellings often shift the hero-centric narrative; for example, Chitra Banerjee Divakaruni's *Bildungsroman* novel, *The Forest of*

Enchantments (2019) weaves a tapestry of narratives from Sita's perspective, questioning blind reverence and devotion towards Ram, whereas Urmila from Kavita Kane's *Sita's Sister* (2014) questions Lakshman's absence from his duties as a husband: 'Does the man have no duty toward his wife and mother?' (Kane 2014: 104). In consequence, these feminist retellings also provide space to question the divine, heroic masculinities of Ram and Lakshman.

While the resurgence of new retellings in post-2000 was initially focused on reclaiming the voices of women and their experiences, we note that increasingly more retellings of the *Ramayana* foreground marginalized masculinities. For example, Amruta Patil's *Sauptik: Blood and Flowers* (2016), Anand Neelakantan's *Asura: The Tale of the Vanquished* (2012) and Ashok Banker's *Ramayana Series: Prince of Ayodhya* (2003) explore the nuances and vulnerabilities that lie beneath the unrealistic extravagance of divine masculinities. These texts position themselves in direct conversation with representations of masculinities in the postcolonial versions, particularly Ram's idealized masculinity.

'Abhi Ram toh Rambo Nikla'[1]: Changing Representations of Ram's and Lakshman's Masculinities

Twentieth-century narratives of the *Ramayana* and their perpetual cultural associations in contemporary India exercise significant influence on the mass cultural consciousness through a variety of media such as films, visual art, comic books, calendars and serials. Banerjee (2011) argues that the emergence of the *Ram Janmabhoomi* movement and its coincidence with the television broadcast of the epic by Ramanand Sagar appealed to the mass middle-class audience for its stable and accessible visualization of the *Ramayana,* which was further reinforced by the comic book *Amar Chitra Katha,* lending a homogeneous iconography of Ram (2011: 115–19). We therefore observe that these all-India sources, together with local and individual modes of retelling the *Ramayana*, serve as the inventory of contemporary visualizations of Ram and Lakshman, and interpretations of their roles as men and symbols. We consider these contemporary narratives the main points of reference for our reading of Arni's and Saket's novels.

Importantly, these representations of Ram's masculinity are significantly different from pre-modern characterizations. Anuradha Kapur (1993: 77–8) notes that while pre-modern retellings of the *Ramayana* already participated in redefining Ram's masculinity and shifting gender dynamics in favour of men, they still did not portray him as hypermasculine, or even masculine by contemporary standards. For instance, Bhavabhuti's early eighth-century version presents Ram as a human who 'behaves like any mortal lover who must be reminded of his duties' (Kapur 1993: 79). Kapur further explains, 'Bhavabhuti's Ram is not the single-minded, unflinchingly determined figure of today's oleographs. Rather he is one assailed by doubts, and self-reproach, one who needs support and who enacts what might appear to be singularly unmanly gestures in today's scale of manliness' (1993: 79). In Tulsi's *Vinayapatrika*, where Ram would be played by boy actors, his divine body is depicted as that of a boy, and not of a muscular man. It is understood that, as a god, Ram does not require a hypermasculine physique, since his powers derive from his divinity. Therefore, though often portrayed as androgynous, Ram remains a deity fully capable of defeating Ravana in combat.[2]

A most significant transformation emerged in the late-twentieth-century iconography of India, with Ram reimagined as a *rashtra purush* – an ideal male embodiment of the nation – and Lakshman constructed as the *deshabhakta* – a devotee of Ram as an embodiment of the nation (Brosius 2004: 100). This was largely facilitated, according to Brosius (2004: 102), by the success of the Doordarshan TV adaptation of the *Ramayana* and the popular *Amar Chitra Katha* series; at the time, the Doordarshan *Ramayan* (1987) was seen by one-eighth of the population of India (Cusack 2012: 279). Arvind

Rajagopal (2004) describes the popularity of the Doordarshan *Ramayan* as follows:

> It drew huge crowds every week. Audience research estimates, compiled by market research companies, climbed from 40 million to 80 million viewers per week over a few months. City streets and marketplaces were empty on Sunday mornings. Events advertised for Sundays were careful to mention: 'To be held after Ramayan'. Crowds gathered around every wayside television set, though few could have seen much on the small black and white sets with so many present. Engine drivers were reported to depart from their schedules, stopping their trains at stations en route if necessary, in order to watch. All of this represented something more than a merely popular TV show. The Rama story belonged to the public domain. (2004: 83–4)

As a result, Ramanand Sagar's (1987) version of Valmiki's *Ramayana* became the most recent way to establish a hegemonic version – a desexualized *Ramayana* that attests to national identity (Pollock 1993: 289; Thapar 2014: 217). Furthermore, Romila Thapar's (2014: 216) study of hegemonic renderings of the *Ramayana* points out that conservative formations have used their own narrow retellings to minoritize and control women and other religions for their political purposes or gains: for a culturally homogeneous nation defined by a self-styled Hinduism.

The twentieth-century mass production of Ram as the 'masculine, ideal persona' created a model of masculinity that chooses to sacrifice domestic responsibilities for the nation, catering to a religion-bound moral framework of preserving the land, women and 'purity' (Thapar 2014: 222). Media such as the comic, which were transformed into a 'cultural commodity' to retell the epics, reaching the mass audience (Singh 2022: 203–4), would instil such radical pedagogies of ideal. These stories, derived from myths, epics, folktales and legends, mass-produced icons of gods and heroes, chiselled and 'sanitized' of eroticism, sexuality and 'vulgarities' to create a network of icons that symbolize the ideal of machismo that is in tandem with the characteristics of an ideal nation and 'Indian identity'. In her study of the comic books *Amar Chitra Katha,* which was 'developed at a moment, when the "Nehruvian" consensus was in crisis', Deepa Sreenivas (2010: 4) points to the manner in which its representations sought 'to train future citizens of the nation through narratives that centre and foreground an indomitable and persevering masculinity' (2010: 4). As such, *Amar Chitra Katha* played a crucial role in the long historical process of redrawing Ram's ideal masculinity.

Varsha Singh (2022) also argues that, immediately after *Amar Chitra Katha*, the iconography of the characters of the Indian epics was further adapted to serve political and religious ambitions, thickening the stereotypes rather than encouraging radical change. For instance, the figure of Ram was transformed from a righteous man to a warrior God of the Hindus – a political orchestration to propel their campaign to claim *Ram Janmabhoomi* (Singh 2022: 207). The Babri Masjid destruction in 1992 and the campaign for the construction of a Ram Mandir in Ayodhya marked a pivotal moment for the representation of Ram. Kapur (1993) observes that Ram's *shanta* was then replaced by anger, rendering him into 'an *ugra* Ram, responding, as it were, to the specific moment, the loss of the janmabhoomi and involved in a fight to regain it' (1993: 103). This resulted in an unusual combination of two models of masculinity: on the one hand, the Hindu meditative masculinity of non-violence (unless a violent reaction is provoked), and on the other hand, the European militaristic models of masculinity, reintegrated through what Brosius (2004) considers 'internalized Orientalism' (2004: 122).

In *Make Me a Man!* (2005), Sikata Banerjee identifies this model of masculinity as 'a warrior-monk', wherein the figure of Ram takes on a militaristic hypermasculinity and is shown as necessary for defending the nation and, implicitly, remaking it into a monolithic culture (2005: 91). The narratives referenced by Banerjee depict Ram's body as muscular, extending his symbolic power by having him '[tower] over a Hindu temple'. This representation resonates with an observation by E. Dawson Varughese (2019) in her visual reading of the book covers of Amish Tripathi's best-selling mythology-inspired fiction (published by Westland):

> Set against dark greys and moody blue hues, a silver (and copper) trishul glints, perfectly placed in symmetry as Lord Shiva's spine follows its middle staff while the blades to the left and right visually echo the shoulder blades of Lord Shiva's body. The image of the trishul is brought into further relief by the shading and shadowing of Lord Shiva's body, revealing a muscular frame and scarring on his left arm and right shoulder. (Dawson Varughese 2019: 177)

Reading these visual tropes, which we may assume are crucial to the marketing of Tripathi's novels, Dawson Varughese (2019) notes that 'Amish's novels are not a strict retelling of epic tales per se' (2019: 175), but both the novels and their covers retain a connection to *itihasa* while gesturing towards the possibility of *darśan*, though unrealized. This reading illustrates the intricate relations that link mythology and religion with popular culture, politics and marketing strategies in post-millennial India, central to which are depictions of Hindu gods such as Shiv and Ram. Depicted as hypermasculine and muscular, Ram is reimagined as the warrior God, exemplifying a broader trend in the current political and cultural reinvention of Ram as an embodiment of the nation-state, capable of defence against any perceived threat. This nationalist framing borrows from the power struggles known from the *Ramayana* itself, idealizing the victory and rule of a unified, homogeneous nation.

De-politicizing Hegemonic Masculinity in Samhita Arni's *The Missing Queen*

The Missing Queen was released in 2013, two years after Samhita Arni published *Sita's Ramayana* (2011), a graphic novel retelling the *Ramayana* from Sita's perspective. Framed as a political thriller, we read *The Missing Queen* as a feminist deconstruction of the politicization of Ram's masculinity, situated within a political reality reminiscent of today's India. The novel itself is set in the aftermath of the Lankan war and Ram's triumphant return to Ayodhya. Its protagonist, an investigative female journalist, embarks on a quest to uncover the truth about Sita's disappearance – a truth suppressed by a state apparatus led by the Washerman, the chief of the secret police. Arrested for confronting Ram on live television with questions about Sita, she meets, in prison, a terrorist from the Lanka Liberation Front (LLF). During an LLF attack on the prison, the journalist escapes and travels to Lanka, where she witnesses the devastation caused by the Lankan war and the true toll the war has taken on the local population. We hence argue that Arni's novel converses with the hegemonic versions of the *Ramayana* as known from the popular culture and politics of postcolonial and contemporary India, offering a multifocal contemplation on the current state of nationalist and masculinist politics, and the political utilization of the *Ramayana* today. In the novel, 'Ayodhya is gleaming, Ayodhya is shining, Ayodhya rivals the heavens, Ayodhya is rich' (Arni 2013: 41), thus reflecting 'New India', where 'Ayodhya Shining' is a reinvention of the political slogan 'India Shining', used by the Bharatiya Janata Party (BJP) in 2004 in a bid to win the parliamentary election. Even though the BJP lost the election (Kaur 2020: 124–5), in Arni's novel, its projection of Ayodhya Shining lives on as a backdrop for her retelling.

Arni focuses on a feminist agenda that interrogates the patriarchal marginalization of Sita and, by extension, the marginalization of women in the current political landscape. In its critique of the political mobilization of idealized Ram, the novel also foregrounds overlooked perspectives, such as Ram's morally questionable actions during the war against Ravana, acknowledging the fact that 'the victors write history [but also wondering] what it looks like to the losers' (2013: 31). By destabilizing the hold that popular retellings of the *Ramayana* have on the national imagination – and their deployment within contemporary nationalist politics – Arni enables readers to critically engage with the notions of masculinity that have come to define Ram in recent decades. At the same time, she invites a more humanized and relatable interpretation of his character. Therefore, we argue that the novel highlights, to further its political aims, a widening gap between the idealized and highly politicized image of Ram's masculinity as a warrior God within recent political discourses and the lived reality of Ram as a man, known from pre-modern *Ramayanas*.

First and foremost, Arni emphasizes the postmodern character of her novel, particularly its focus on the plurality of competing narratives surrounding the story of Ram and Sita, as well as their power dynamics. According to the official narrative, 'This is a war that has been justified by the concept of duty. Ram's duty, as a husband, was to rescue Sita' (2013: 37). Yet the son of Vali, Angada, highlights a different perspective, reminding the readers that 'Ram killed [his] father, he shot him in the back, it was a dishonourable thing to do' (2013: 35). In the official version, however, 'every injustice, every wrong that has been committed, has been changed into right' (2013: 38), with Kaikeyi reflecting:

> As for Ram, he is a living god to the people of Ayodhya. He holds himself to an impossible standard (…). But he can't see beyond himself – he's obsessed with his actions, with his nature (…) striving to be the ideal. That can be another form of cruelty. He was cruel to his father, mother, me, his brothers, and his wife. (2013: 6)

As has been noted, earlier forms of the *Ramayana* variously represent Ram, providing diverse interpretations of Ram's complex role as a *Kshatriya* but also, at least partially, an ascetic. Seema Bawa (2018) writes that in the *Aryanakanda* of Valmiki's *Ramayana*, Surpanakha comments on 'this ascetic-warrior contradiction' (2018: 118) in Ram: 'The appearance then is a physical marker of the acceptance of the exile, the renunciation of worldly luxuries, the life of princes, living in forests like ṛṣis, foraging for food. However, the retention of weapons, symbols of kṣatriya virility, signifies their ability to protect, fight, survives [*sic*]' (2018: 118). Arni also renders Ram a man who feels and suffers, echoing Valmiki's *Ramayana* (Doniger 2010: 222), especially as he struggles with emotional conflicts and traumas after the tragic Lankan war and Sita's abandonment by Lakshman in the forest. The first display of emotions occurs during an interview with Ram, who, '[a]gitated, runs his hands through his hair [saying:] "Well ... sometimes ... It had been difficult. Those years were so difficult – exile, war, coming back"' (Arni 2013: 26).

This extends to other descriptions of Ram's emotionality, which appear in the minor narratives of the epic war: 'I saw Ram, standing a few steps behind Sita. He looked angry and confused. He could hear her crying, but he couldn't understand it. She then turned towards him, and I saw revulsion flit across his face. Revulsion and anger' (2013: 52). Particularly, the moment of Ram and Sita's reunification is presented in a way that contrasts with idealizing narratives. In Arni's version, Ram is shown as internally torn between his duties as king and husband, and the resentment he holds for Sita: 'I suppose he must have been confused. He was angry – all this war. (...) She isn't the woman he remembers. She bursts into tears. He's fought for her – and she's crying' (2013: 92). With 'Sita [being] the price that Ram has, unwittingly, paid to rule an empire' (2013: 123), Ram is eventually left with a fraught deal: '[i]n Ayodhya, (...) people (...) call it a great sacrifice, the sacrifice a king must make to fulfil the promise of his birth, and his duty to his people. Ayodhya venerates Ram for this' (2013: 123). Unlike the now hegemonic narratives and visual depictions, Arni's representation of Ram foregrounds his emotional experience of anger, jealousy and disappointment, which underscores the injustices that Sita endured at Ram's hands, as well as humanizing Ram and his display of 'imperfect' masculinity in the process.[3]

This stands in contrast to the politicized image of Ram as the 'ideal' man that dominates popular discourse as a symbol. Arni captures this politicization by depicting Ayodhya's 'walls and billboards (...) [as] plastered with images of Ram' (2013: 23). This ubiquity of Ram's imagery resonates with the nationalist fervour to promote Ram's image in public life, particularly through the Ram Mandir campaign. However, in Arni's book, Ram's construction differs from what scholarship on dominant retellings of the *Ramayana* have typically shown. Instead of a hypermasculine, muscled Ram, we read in the *Ayodhya Times* that: 'We have a god-like king, Ram, the embodiment of virtue, gracing the throne of Ayodhya. A king who rules with disinterest, with no thought of himself. An ideal king – the philosopher-king' (2013: 39). Interestingly, Ram is not shown as a warrior or a warrior-monk; his uniqueness is found in the divine aura of a 'philosopher-king'.[4]

In his critique of hegemonic masculinity, Michael Moller (2007) pointed out that Connell based her framework on a critique of the positivist tradition, 'for the way it seeks, and inevitably finds, self-contained and deterministic patterns within men's lives, a pattern which is always already anticipated as 'masculinity'" (2007: 267). However, Connell's own 'conceptual work on masculinities, especially the notion of a hegemonic form of masculinity, also [appears to articulate] a desire for such orderly patterns' (Moller 2007: 267), which overlooks nuanced power dynamics and actual experiences of masculinity subjects. We therefore argue that *The Missing Queen* subscribes to and illustrates Moller's critique through its construction of the male characters as conflicted figures, rather than as one-dimensional embodiments of preconceived notions or mere symbols, however potent their influence may be on the lived realities of many men in today's India.

Arni situates Ram within what Susan Bordo (2000: 242) describes as 'the double bind of masculinity' – a setting where men and boys are subjected to mutually contradictory expectations. Bordo comments: 'The fact that these contradictory directives put a *real* person in a difficult (if not impossible) double bind gets masked by the fact that we've created numerous *fictional* heroes who successfully embody both requirements' (2000: 242). Arni's Ram suffers under his own double bind – that of a devoted philosopher-king, on the one hand, and the perfect husband, on the other. Showing Ram as struggling to satisfy the impossible expectations, Arni de-fictionalizes him, rendering him instead as a real man, whereas his idealized masculinity is exposed as a media-constructed, rigid set of attributes and virtues that ultimately fail to secure actual power for him while constraining his ability to live his life freely.[5] Through this depiction, Arni deconstructs Ram's masculinity as a fictional instrument for the potential disciplining and 'double-binding' of other real men.

This deconstructive process of humanization encompasses both brothers in the novel and unfolds as a corporeal procedure as well. First, Arni de-idealizes the bodies of Ram and, especially, Lakshman. Rather than depicting them as perfectly chiselled, eternally youthful bodies, they both succumb to ageing, with 'the close-ups betray[ing Ram's] age, magnifying the faint streaks of grey in his hair, the crow's feet at the corner of his eyes' (Arni 2013: 12). The passage of time seems to leave an even bigger mark on Lakshman, who 'was the most handsome of Dasaratha's sons, once upon a time. Even after the return from Lanka, he outshone Ram. But since then, it's been a decline. His eyes are puffy, his large belly strains against his general's uniform' (Arni 2013: 23). He is presented as an ageing man, whose physique increasingly differs from the renderings of him as an agile and formidable warrior; now he is 'pot-bellied, sweating, anxious' (Arni 2013: 67). In representing him this way, Arni closes the semantic gap between the brothers as carriers of symbolic patriarchal power and their reality as embodied, fallible men, thus echoing Bordo's (2000: 84–104) work on the disjunction between phallic conceptions of patriarchal masculinity and the physical vulnerability of the male body.

In consequence, Arni's feminist retelling of the epic serves to illustrate the tensions that underpin the social patterns of hegemonic and idealized masculinities and their incongruity with the lived experiences of those who embody them. In the final sections of the novel, Kambhan reflects: 'It's not Ram, really, who wronged her [Sita] the most. We did. Ayodhya did. We didn't believe her. We didn't trust her. We forced her out' (Arni 2013: 130). With this conclusion, *The Missing Queen* reframes the discussion of masculinity, men's duties and women's oppression in the *Ramayana*'s now popular retellings from an individual-focused parable to a broader contemplation of the social conditioning of gender models and their consequences for lived realities. The novel engages with the politicized developments in Ram's iconography as a hypermasculine and valiant figure, juxtaposing cultural projections of hegemonic, idealized masculinity and the related social expectations with the lived realities of subjects who appear to embody these masculinity models. Arni partially enacts a critique of the framework of hegemonic masculinity by revealing the inability of such models to accommodate diverse, individualized experiences of men who are socially and culturally assumed to embody hegemonic masculinity. Specifically, as a feminist retelling, *The Missing Queen* critiques the patriarchal bias in the currently popular versions of the *Ramayana*, while also questioning narratives of idealized masculinity and unproblematic male power. Arni demonstrates that such masculinity models fail to express emotionality and the fallibility of male bodies, along with other non-ideal aspects of masculinity that are conspicuously absent from socially recognized and culturally enforced models of exemplary traditional Indian masculinity.

Deconstructing Idealized Brotherhood in Pervin Saket's *Urmila*

Our reading of Saket's feminist retelling of the *Ramayana*, titled *Urmila* (2016), focuses on subverting the idealization of cultural representations of the brotherhood of Ram and Lakshman, particularly Lakshman's sacrifice, unwavering loyalty and devotion. The construction of Lakshman appears to be rooted in traditional models of masculinity associated with the postcolonial project of nation-building. His representation as a *(desh)bhakta,* or devotee who surrenders to the divine, is used to embody hyper-nationalist followers who conflate devotion with political loyalty. In contrast, Urmila's narrative, and especially her tragic love for Lakshman, has been overshadowed by the idealization of Lakshman and the discursive emphasis on his fraternal duties towards Ram. We argue that retellings like *Urmila* (2016) offer a critique of cultural understandings that idealize normative masculinity, such as that of Lakshman, intrinsically tethered to a brotherhood that romanticizes sacrifice and fraternal duties, at the cost of reinforcing domestic inequalities.

When Ram is banished from Ayodhya in the *Ramayana*, Lakshman abandons his marital duties to follow his brother and sister-in-law, Sita, into exile. Popular narrative traditions of the *Ramayana* idealize and celebrate Lakshman's unwavering devotion to his brother. For instance, Ramanand Sagar's representation of Lakshman as the 'ideal' brother functioned as an appeal to the traditional moral framework for 'ideal' masculinity, owing to its deep ties to family, sacrifice and devotion (Thapar 1991: 145–51). Set in today's India, *Urmila* follows the story of Urmila Karmakar (named after Urmila from the *Ramayana*), a young woman living in urban Mumbai, who feels trapped in a 'loveless' (Saket 2016: 19) marriage with Shree (a reinvention of Lakshman), a man from a wealthy and well-affiliated family. This narrative reimagines the lesser-known character of Lakshman's wife and explores her emotional complexities by interweaving traditional epics and contemporary reflection. In this creative parallel to Lakshman following Ram and Sita into a fourteen-year exile in the *Ramayana*, Shree leaves behind his wife, Urmila, to follow his brother, Purushottam (Ram) and his wife, Vanita (Sita), to Dubai. This narrative, through Urmila's point of view, explores themes of sacrifice, identity, solitude and emotional isolation, capturing Urmila's psychological journey within the constraints of an 'idealized' masculinity and the prolonged absence of her husband.

This feminist rewriting of the *Ramayana*, with its centralization of domestic affairs, reclaims Urmila's silence, which, in Valmiki's *Ramayana*, is depicted as a sacrificial act wherein she agrees to sleep on Lakshman's behalf so he can guard Ram and Sita day and night. In Valmiki's *Ramayana,* the focus is on Urmila's silence during her fourteen years of sleep, which, as a consequence, marginalizes and silences her story, even as her sacrifice enables her husband's fulfilment of his duties as a devoted brother and protector, a *bhakta* (Tiwari and Chaudhary 2021: 6). Urmila, in Saket's rendition of the *Ramayana,* however, is anything but silent. Traversing the first-person narration of her disappointment with the 'loveless marriage', we encounter the complexities of Urmila's emotions and her utilization of art to understand the repercussions of traditional, idealized masculinity for women. Foregrounding the cost borne by women, the novel reveals 'fault lines' within the archetype of the ideal, loyal and sacrificing brother – Lakshman. This, therefore, subverts the postcolonial reverence for such masculinity, bringing attention to its unacknowledged consequences. Among these, we find the 'absent man' motif particularly interesting, as it questions the unrealistic idealizations of the virtuous, devoted brother but also, importantly, situates this discussion in the real-life context of India's Gulf migration realities.

While the currently dominant narratives of the *Ramayana* are mostly preoccupied with Ram as an embodiment of virtue and the 'ideal' man (*maryada purushottam*), his character, along with the traditional masculinity he embodies, is incomplete without Lakshman – the obedient, protective and loyal brother. While Ram, as we suggest, symbolizes the 'ideal' man and also the 'ideal' nation, without Lakshman's unyielding fidelity and self-sacrifice, Ram, as such, would not exist. The duality of this idealized masculinity is contemporarily reflected in cinematic representations of nationhood through brotherhood. Bombay cinema, Jyotika Virdi argues (2003: 87), 'narrows the gap between the public and private' by treating the family as emblematic of the nation. Bollywood, from the 1970s masala films to contemporary productions, often portrays an almost devotional fidelity between brothers. Drawing on the epics like the *Ramayana*, films such as

Deewar (1975), *Ram Lakhan* (1989), *Amar Akbar Anthony* (1977), *Naam* (1986), *Karan Arjun* (1995), *Hum Saath-Saath Hain* (1999), *Main Hoon Na* (2004) and others have romanticized brotherhood, intertwining masculinity with a sense of morality and virtue that demands personal sacrifice, including detachment from marital duties. In other words, Saket's retelling of the *Ramayana* de-idealizes this image of masculinity and brotherhood. By re-reading the 'perfect man' as embodied by Purushottam and, more importantly, Shree, this novel questions the long-standing hero/god-like traditional masculinity and its unrealistic standards. Stripped of its grandeur, Lakshman's/Shree's masculinities seem incapable of offering emotional intimacy or personal care, aspects generally considered imperative to marital and domestic life.

Written over seven chapters, the novel is distinguished by the seven vows the bride and groom promise to uphold through their marriage. According to Ruth Vanita (2005), in the Hindu wedding rite, the *saptapadi* (seven steps), the bride and groom perform seven circumambulations around the sacred fire, during which they recite seven verses that exalt a particular dimension of marriage (2005: 128–9). One such vow in the novel reads: 'Dhve Urjve – Vishnuthva – Anvethu. With the second vow of Saptapadi, as your husband, I promise you Urje: to grow strong together, physically, mentally, and spiritually' (Saket 2016: 39). These vows are soon broken by Shree, whose emotional absence first manifests in his disinterest in sex, including consummating the marriage. Urmila recounts several episodes of such disinterest:

> Naked, because his gaze did not cover my body. In the omniscient glare, I was muscle, tissue, and skin, as I lay on a surgeon's table. My curves were lost, the fluttering in my heart dismissed, the goosebumps on my skin ignored. (Saket 2016: 98)

Initially misconstruing this disinterest as the awkward nature of his personality, Urmila later realizes that it is his obsession with maintaining appearances of the 'ideal' brother and son that leaves no space for an outgoing, loud and bright-clothed Urmila in his life:

> I realised soon that my banter put him off. He liked me receded and ebbed. He liked me as a backdrop, beautiful and unobtrusive. My sight was too loud, my clothes too bright, my hair untamed, my laughter too quick. I had to learn restraint. (2016: 98)

We suggest that such desexualization of men might echo the construction of a homogenized masculine identity during the nationalist struggle for Indian independence, which emphasized the de-eroticization of Indian male sexuality (Alter 2011: 24–37; Dasgupta and Gokulsing 2013: 9) in promoting Gandhi-like masculinities, along with celibacy and semen retention, rather than physical strength (Alter 1996; 2011). Thus, Shree's rejection of Urmila's body signifies a masculinity that thrives on sexual restraint. For instance, Urmila's futile attempts to form an intimate relationship with Shree lead her to remark that Shree:

> was the happiest with me when:
> I cooked Sabudana khichdi (Puru's favourite, not his).
> I massaged his father's legs,
> I took over supervising the maid.
> I relieved his sister of all her chores.
> I chatted over tea with his mother.
> I wore cotton saris,
> I didn't wear bangles,
> I spoke less in company at the dining table, and I fell asleep immediately after he entered our room. (2016: 99–100)

From this list of Urmila's observations, Shree's approval of her traditionally domestic demeanour, depicted by her selfless servitude, reflects narratives of a controlled gendered identity that is devoid of sexuality, desires and individuality. Shree's lack of individuality, as a result of his obsession with his brother, reveals that his 'ideal wife' is a woman in the background – 'beautiful and unobtrusive' (Saket 2016: 98).

Furthermore, Shree shows a strong sense of responsibility for his sister-in-law, Vani (Urmila's adoptive sister and sister-in-law, who parallels the character of Sita). His admiration for Vani seems embedded in his belief in the ideals of a traditional Indian wife – in understanding 'where a woman's dignity lies' (2016: 208). Shree's admiration for Vani also stems from their shared obsession with and admiration for Purushottam. Urmila says:

> Shree's conviction about Vani's virtues was based on an irrefutable logic – she was attached to Puru and hence embodied an extension of all his goodness. When Ravi had spilled the beans on the relationship between Vani and Puru, Shree did not care about its inappropriateness. He only registered that Puru had chosen her, and so she was blameless. (Saket 2016: 207)

Hence, the more Vani floated in his realm of 'goodness' (Saket 2016: 207), the more Urmila was rejected from the triad of Shree, Vani and Purushottam, where Purushottam was the object of desire and worship.

As mentioned earlier, in parallel with Ram's divine ideals and virtues, Purushottam embodies not only the 'ideal' man but also the ideal Indian nation. The 'ideal' Indian man practices restraint, self-discipline and passivity, and holds a strong sense of protecting the nation (Alter 1996: 305; 2011: 7). It demands sacrifices of the family, sexual desires, and self-centredness to become an ideal citizen of the nation. In Saket's retelling, Shree's strong sense of his fraternal duty to protect Purushottam and Vani leads him to abandon Urmila and his parents, bringing into question the detachment and abandonment of domestic responsibilities that come along with performing his role as the 'ideal' brother and *bhakta*. Shree, before fleeing to Dubai, tells Urmila, 'I cannot stay here anymore. Where there is no place for Dada, there is no place for me' (Saket 2016: 203). Adorned with symbols of marriage (mangal sutra, bindi, sindoor, bangles), Urmila lives in a loveless marriage without a husband, left by the 'fleeting sons' to a land that was their 'mistress' (Saket 2016: 37).

This breakdown of marriage due to Shree's absence serves a dual function, as we argue. It challenges the idealization of Lakshman's loyalty, which has been repurposed to define male duty, while contemporizing the *Ramayana* and its lessons of masculinity and matrimony by anchoring the narrative in the lived reality of Gulf migration. By thus reinventing the story of Urmila and Lakshman, Saket de-privatizes the Gulf experience in the context where 'while the Gulf is much spoken about, it is spoken about as if it is a private affair' (Karinkurayil 2024: 14). Similarly, Saket renegotiates this tension, transforming the private yet common experience of migration and its effects on the family into a subject of public discourse through literature, while re-domesticating the construction of Lakshman's masculinity as the ideal 'national' *bhakta*.

Although migration to the Gulf is frequently framed through economic factors, as Gayathri Prabhu (2023) notes, importantly, it more often than not comes at a significant emotional cost: 'loneliness, secrecy, shame, deep-seated rage (…), [with much less attention paid to] a murky, underground miasma of what is equally unspoken in both high policy discourse and in the intimate bedrooms and inner courtyards of families' (2023: 25). This secrecy, as Saket shows through the example of Urmila, disproportionately affects women, particularly those left behind, who 'actively support the migration of their husbands and brothers through their own labor, although this work often goes unacknowledged. The femininity of migrants' wives becomes reinforced by their role in helping fulfil their husbands' duties of care' (Wright 2020: 454). Urmila, however, refuses to confine herself to the passive role of the rejected, helpless wife and decides to fulfil her desire for children without her husband. While the entire narrative speaks of the (in)visible Shree and his neglect, rejection and absence, his presence is made known to the readers through her reminiscent thoughts, artwork, and her pursuit of children. Urmila's decision to undergo 'artificially assisted pregnancy' (Saket 2016: 262) represents a finality in her fourteen years of sorrow and longing. It signals her empowerment as a self-reliant woman while simultaneously challenging the cultural ideal of Shree, and by extension Lakshman and the husband in general, as paragons of virtue and unquestionable agents of patriarchal authority.

Pervin Saket's *Urmila* thus recreates the twentieth- and twenty-first-century popular versions of the *Ramayana* through a feminist lens, creating space for counter-narratives that deconstruct the culturally reinforced portrayals of loyalty, devotion, brotherhood and sacrifice. This novel reflects constructions of a performative model of an idealized masculinity that suppresses emotions to uphold an unwavering fraternal devotion to a traditional patriarchal order, embodied by Ram. While narrow retellings of the nationalist movement interpret the *Ramayana* to revere Lakshman for his unwavering loyalty, Saket's contemporary take on him as a man and a husband reveals the unrealistic standards for loyalty that ultimately translate into rejection and abandonment. Such narratives thus foreground features from other competing versions of the *Ramayana*, obscured by the dominant versions, which reinforce the nationalist discourses of masculine power, national *seva*, and repressed emotionality. Rao's (1991) study of women's folk songs in Andhra Pradesh unearths the subversive interpretations of the *Ramayana* that add vulnerability to Lakshman's militarized selfhood, thus questioning cultural notions around familial duties beyond being an ideal brother to Ram. Rao posits that these women sing of Lakshman as a surrogate father taking care of Sita's sons; he notes, 'He puts oil on their scalps, feeds them milk, and they urinate on his clothes. Lakshmana loves it; his face glows like the full moon' (1991: 132). These feminist folk songs rewrite nationalist ideology's symbolism, shifting the focus to Lakshman's representation as a householder, instead of his valorized embodiment as the emotionally restrained, sacrificing brother who serves and executes Ram's moral law. Similarly, *Urmila*'s feminist intervention into the *Ramayana* urges a masculinity for Lakshman that combines emotional complexity and marital duty.

Conclusion

By tracing representations of 'idealized' masculinity in Rama and Lakshman within postcolonial and post-millennial reinterpretations of the *Ramayana*, we have established that contemporary feminist retellings of the *Ramayana* reshape the archetypes of idealized masculinities in lived realities. These retellings, we have argued, create a foundation for exploring subversions that allow marginalized voices and discursive spaces to de-idealize and humanize the deified masculinities of the twentieth- and twenty-first-century versions of the *Ramayana*. Samhita Arni's *The Missing Queen* and Pervin Saket's *Urmila* challenge the postcolonial cultural constructions of the 'perfect man,' which enforce emotional restraint and unquestioning loyalty towards one's *dharmic* (often national) duty. This chapter also calls into question the complementary nature of the postcolonial constructions of Ram and Lakshman – Ram as the *maryada purushottam* and Lakshman as the idealized moral adjunct to Ram. While Ram and Lakshman are presented as incomplete without one another, Arni and Saket flesh out the nuances of the two's romanticized brotherhood.

Our reading suggests that the concept of hegemonic masculinity is inadequate to account for Ram's and Lakshman's gender constructions, given their political and cultural idealization. Both novels reveal, albeit indirectly, that reading certain male figures as embodiments of idealized virtues renders them fictional and disconnected from the lived realities of men, even those men who appear to benefit most from their patriarchal privilege. In this sense, the novels' critique of such masculinity-formation subscribes to Moller's critique of hegemonic masculinity as an unrealistic category, which, on its own, fails to accommodate the complexity of lived male experience. Interestingly, however, Ram and Lakshman's culturally privileged positions within India allow them a hegemonic influence over social, religious and cultural expectations around masculinity, as their hegemony is not dependent on transient ideals. Unlike in Connell's framework, their hegemonic masculinities are not a combined result of the *Zeitgeist* and the power hierarchies of the moment; therefore, they are uniquely positioned to serve as agents of change, endorsing alternative models of masculinity when discursively reclaimed in feminist fiction.

Invoking representations of Ram and Lakshman from classical texts as emotionally complex, we suggest that both novels illustrate how nationalist discourses have reinvented the brothers, particularly Ram, as emotionally suppressed symbols of an idealized masculinity. The analysed novels (re-)write Ram and Lakshman not as divine embodiments of virtue but as complex, emotionally vulnerable and flawed humans. This shift in representing Ram and Lakshman as god-like heroes and warriors to more humanized, flawed versions of their characters creates alternative realities that are rooted in silent dissent, grief, anger and pain. Arni and Saket thus expose the 'perfect man' as one that is essentially restrictive and dehumanizing, while gesturing towards more relatable models of masculinity.

Notes

1 In *Main Hoon Na* (2004), one of the characters exclaims 'Abhi Ram toh Rambo nikla' – meaning, 'Ram has turned out to be Rambo' – when the hero, Ram (played by Shah Rukh Khan), reveals his true identity as a fit-bodied major in the Indian army during a daring rescue of his long-missing brother, Lucky (Lakshman).

2 Interestingly, Kapur also writes that 'even in the midst of battle, Ram is always composed, always tranquil, never angry. The angry, "human" response is left to Lakshman, whose exuberances are proverbial' (1993: 88).

3 Arni's portrayal of Lakshman reflects similar emotional complexity, as he struggles with the trauma of war and the consequences of his brother's cruel decisions, dictated by societal expectations. At one point, we read: 'When Lakshman came back, he was a changed man. He was never the same. Always racked by guilt' (2013: 122). The

trauma of the events that followed Sita's brief return to Ayodhya 'eats away at Ram, it plagues Lakshman with nightmares' (2013: 123).

4 With his semi-divine status in *The Missing Queen*, Ram is portrayed as a god-like politician, who, however, does not claim divinity himself; it is ascribed to him through calculated propaganda, making him 'an intimate presence in [the Ayodhyans'] living rooms' (2013: 11). Therefore, Ram's portrayal in *The Missing Queen* diverges from the militaristic, hypermasculine representations. Instead, those qualities are delegated to Hanuman, whose '[m]uscles threaten to burst the stems of his tight-fitting uniform. His simian features, beloved and so familiar, decorate many a young boy's bedroom and his face adorns army recruitment posters' (2013: 114). By extracting the militarism associated with Ram's embodiment as the 'ideal' Indian man and attributing it to his loyal *bhakta* Hanuman, Arni reserves for Ram an idealized form of noble masculinity.

5 While the true power remains in the hands of Hanuman and is soon to be transferred to the people through Ayodhya's democratization, Ram emerges in the novel as a figure whose idealized masculinity wields no actual influence beyond its performative value for the public. This underscores that Ram's meticulously crafted public persona is merely a façade.

References

Alter, Joseph. 1996. 'Gandhi's Body, Gandhi's Truth: Nonviolence and the Biomoral Imperative of Public Health'. *The Journal of Asian Studies* 55(2): 301–22.

Alter, Joseph. 2011. *Moral Materialism: Sex and Masculinity in Modern India*. New Delhi: Penguin Books.

Arni, Samhita. 2013. *The Missing Queen*. New Delhi: Zubaan.

Arni, Samhita and Moyna Chitrakar. 2011. *Sita's Ramayana*. Chennai: Tara Books.

Banerjee, Anindita. 2011. 'A Higher Narrative in Pictures: Iconography, Intermediality, and Contemporary Uses of the Epic in India'. In *Epic and Other Higher Narratives*, edited by Steven Shankman and Amiya Dev, 112–27. Noida: Pearson.

Banerjee, Sikata. 2005. *Make Me a Man! Masculinity, Hinduism, and Nationalism in India*. Albany, NY: State University of New York Press.

Banker, Ashok. 2003. *Ramayana Series: Prince of Ayodhya*. New Delhi: Orbit.

Barjatya, Sooraj. 1999. *Hum Saath-Saath Hain*. India: Rajshri Productions.

Bawa, Seema. 2018. 'Visualising the Rāmayāna: Power, Redemption and Emotion in Early Narrative Sculptures (*c*. Fifth to Sixth Centuries CE)'. *Historical Review* 45(1): 92–123.

Bhatt, Mahesh. 1986. *Naam*. India: Aryan Films.

Bordo, Susan. 2000. *The Male Body: A New Look at Men in Public and in Private*. New York: Farrar, Straus and Giroux.

Brosius, Christiane. 2004. *Empowering Visions: The Politics of Representation in Hindu Nationalism*. London: Anthem Press.

Butler, Judith. 1988. 'Performative Acts and Gender Constitution: An Essay in Phenomenology and Feminist Theory'. *Theatre Journal* 40(4): 519–31.

Butler, Judith. 1993. *Bodies That Matter: On the Discursive Limits of 'Sex'*. London: Routledge.

Butler, Judith. 2007. *Gender Trouble*. New York: Routledge.

Chopra, Yash. 1975. *Deewar*. India: Trimurti Films.

Connell, Raewyn W. 1995. *Masculinities*. Cambridge: Polity Press.

Connell, Raewyn W. and James W. Messerschmidt. 2005. 'Hegemonic Masculinity: Rethinking the Concept'. *Gender and Society* 19(6): 829–59.

Cusack, Carole M. 2012. 'The Gods on Television: Ramanand Sagar's *Ramayan*, Politics and Popular Piety in Late Twentieth-Century India'. In *The Handbook of Hyper Real Religions*, edited by Adam Possamai, 279–97. Leiden: Brill.

Dasgupta, Rohit K. and K. Moti Gokulsing. 2013. 'Introduction'. In *Masculinity and Its Challenges in India: Essays on Changing Perceptions*, edited by Rohit K. Dasgupta and K. Moti Gokulsing, 5–25. Jefferson, NC: McFarland.

Dawson Varughese, E. 2019. 'Seeing Shiva, Seeing Ram: Visual Representations of Deities in the Genre of Mythology-Inspired Fiction, An Analysis of the Book Covers of Amish Tripathi's Novels'. *South Asian Popular Culture* 17(2): 171–83.

Desai, Manmohan. 1977. *Amar Akbar Anthony*. India: MKD Films.

Divakaruni, Chitra Banerjee. 2019. *The Forest of Enchantments*. Noida: Harper Collins.

Doniger, Wendy. 2010. *The Hindus: An Alternative History*. Oxford: Oxford University Press.

Ghai, Subhash. 1989. *Ram Lakhan*. India: Mukta Arts and Suneha Arts.

Gokhale, Namita and Malashri Lal (editors). 2009. *In Search of Sita: Revisiting Mythology*. Gurugram: Penguin Books.

Kane, Kavita. 2014. *Sita's Sister*. New Delhi: Rupa Publications.

Kapur, Anuradha. 1993. 'Deity to Crusader: The Changing Iconography of Ram'. In *The Hindus and Others: The Question of Identity in India Today*, edited by Gyanendra Pandey, 74–109. New Delhi: Viking.

Karinkurayil, Mohamed Shafeeq. 2024. *The Gulf Migrant Archives in Kerala*. Online. Oxford: Oxford University Press.

Kaur, Ravinder. 2020. *Brand New Nation: Capitalist Dreams and Nationalist Designs in Twentieth-Century India*. Stanford, CA: Stanford University Press.

Khan, Farah. 2004. *Main Hoon Na*. India: Eros International.

Moller, Michael. 2007. 'Exploiting Patterns: A Critique of Hegemonic Masculinity'. *Journal of Gender Studies* 16(3): 263–76.

Neelakantan, Anand. 2012. *Asura: The Tale of the Vanquished*. New Delhi: Leadstart Publishing.

Patil, Amruta. 2016. *Sauptik: Blood and Flowers*. Noida: HarperCollins.

Peltola, Marja. 2022. *Nuancing Young Masculinities: Helsinki Boys' Intersectional Relationships in New Times*. Helsinki: Helsinki University Press.

Pollock, Sheldon. 1993. 'Rāmāyaṇa and Political Imagination in India'. *The Journal of Asian Studies* 52(2): 261–97.

Prabhu, Gayathri. 2023. 'A Gulf of Secrets: Priya Kuriyan's Graphic Memoir "Ebony and Ivory"'. *The Journal of Commonwealth Literature* 58(1): 22–35.

Rajagopal, Arvind. 2004. *Politics after Television: Religious Nationalism and the Reshaping of the Indian Public*. Cambridge: Cambridge University Press.

Rao, Velcheru Narayan. 1991. 'A Ramayana of Their Own: Women's Oral Tradition in Telegu'. In *Many Ramayanas: The Diversity of a Narrative Tradition in South Asia*, edited by Paula Richman, 114–36. Berkeley, CA: University of California Press.

Richman, Paula (editor). 1991. *Many Ramayanas: The Diversity of a Narrative Traditions in South Asia*. Berkeley, CA: University of California Press.

Richman, Paula. 2001. 'Questioning and Multiplicity within the Ramayana Tradition'. In *Questioning Ramayanas: A South Asian Tradition*, edited by Paula Richman, 1–25. Berkeley, CA: University of California Press.

Roshan, Rakesh. 1995. *Karan Arjun*. India: Film Kraft.

Sagar, Ramanand. 1987. *Ramayan*. India: IST Mumbai.

Saket, Pervin. 2016. *Urmila*. Mumbai: Jaico Publishing House.

Sen, Nabaneeta Dev. 2008. 'Alternative Interpretations of the Ramayana: Views from Below'. In *Women's Studies in India: A Reader*, edited by Mary E. John, 583–91. New Delhi: Penguin Books.

Shulman, David. 1991. 'Fire and Blood: The Testing of Sita in Kampan's Iravamavataram'. In *Many Ramayanas: The Diversity of a Narrative Tradition in South Asia*, edited by Paula Richman, 89–114. Berkeley, CA: University of California Press.

Singh, Varsha. 2022. 'Epics as Cultural Commodities: Comics Books of the *Ramayana* and the *Mahabharata*'. *The Journal of Commonwealth Literature* 57(1): 203–23.

Sreenivas, Deepa. 2010. *Sculpting a Middle Class: History, Masculinity and the Amar Chitra Katha in India*. New Delhi: Routledge.

Thapar, Romila. 1991. 'A Historical Perspective on the Story of Rama'. In *Anatomy of a Confrontation: The Rise of Communal Politics in India*, edited by Sarvepalli Gopal, 141–63. London: Bloomsbury Academic.

Thapar, Romila. 2014. *The Past as Present: Forging Contemporary Identities through History*. New Delhi: Aleph Book.

Tiwari, Aditi and Priyanka Chaudhary. 2021. 'Revisioning Subalternity: A Critical Study of *Ramayana* through Mandavi and Urmila'. *Rupkatha Journal on Interdisciplinary Studies in Humanities* 13(2): 1–9.

Vanita, Ruth. 2005. *Love's Rite: Same-Sex Marriage in India and the West*. New York: Palgrave Macmillan.

Virdi, Jyotika. 2003. *The Cinematic Imagination: Indian Popular Film as Social History*. New Brunswick, NJ: Rutgers University Press.

Woolcock, Joseph A. 1985. 'Politics, Ideology and Hegemony in Gramsci's Theory'. *Social and Economic Studies* 34(3): 199–210.

Wright, Andrea. 2020. 'Making Kin from Gold: Dowry, Gender, and Indian Labor Migration to the Gulf'. *Cultural Anthropology* 35(3): 435–61.

6

Performing and Recalibrating Selves, Space and Memory in '*Measures of Elsewhere*'

Anusha Ravishankar, Carmen Wong and Nikhil Govind

Introduction

'Measures of Elsewhere' is a performance that sketches an encounter between two characters – Insider and Outsider, both seeking to preserve the places, practices, and memories of where they have come from, within the space they currently exist in: the kitchen. We revisit 'Measures' after almost a decade through an auto-ethnographic account of the devising of the performance that draws from Insider's and Outsider's personal narratives and archives. It tracks the reflexive digestion of the memory of various performance leftovers: written reflections, performance notes, scripts/scores, ambient sound and music recordings, rehearsal videos, personal journal entries, photographs and conversations. In the case of some of these archival materials, we offer them as they were photographed, written, or journalled (without our current reflections on them) as totems of times and selves we are unable to fully access today.

This collaboration transpired between two theatre researcher-practitioners, Anusha Ravishankar and Carmen Wong, who are fascinated with transcultural and intercultural performances, contemporary feminist dialectics, (dis)placement and memory, and the adaptation of narratives from a nomadic-postcolonial-diasporic-Asian-female[1] perspective. We began working on 'Measures' in

New Delhi, supported by an Asian Cultural Council grant, in the winter of 2015, half a year after we graduated from the MA programme in International Performance Research, where we first met. A majority of the work we did individually and together as researchers – in class, in the studio and the black box – while at the University of Warwick was rooted in the methodologies of practice-as-research, in which we prioritized a deepening of collaborative friendships, important to our approach because of the authentic exchange, learning, and trust it engenders.

Our combined research inquiry was initially ambitiously framed as follows: How does one respectfully, culturally and intertextually[2] (Hutcheon 2006: 21–2) translate and adapt a staged collage about performing gender, embodying personal memory, and enact symbols within its palimpsestuous echoes? Who exactly are the co-creators of the piece, when the piece is adapted in situ: co-authors, performing devisers, the audience? How and why do they contribute? Looking back now, we feel the need to deepen our inquiries by including the ways in which our research could be embodied. Our post-research inquiries (updated in 2024) include: What stories are our bodies yearning to tell? What desires are behind them? Is there a style or quality of gesture that is critical to the successful telling? Can we use these inquiries to read and perform everyday choreographies, where choreography is 'other than or additional to the arrangement of bodily movement' (Allsop and Lepecki 2008: 1–4) and can be a means to observe and read the body set in the domestic everyday?

Practice as Research (PaR) as a methodology for performance studies scholarship, while still in its developmental stages in India, has benefitted from the work of practitioner-scholars like Anuradha Kapur,[3] Veenapani Chawla, Rustom Bharucha and, more recently, dancers like Ranjana Dave. Using PaR, this chapter demonstrates the practice-based exercises used to collect relevant research data. These exercises are created and enhanced in collaboration with other practitioners and scholars of theatre and performance studies. The findings make their way into the final live performance and the chapter carefully charts this journey for the reader. While we focus on the details of the research process, we also find it crucial to highlight the affective implications of embodied research. Here, we gain from the act of looking back at our respective archives after a long pause, and choosing to use material and texts personal and private. This also enables us to introduce an intervention in the form of academic writing, interspersing it with our reflections, performance scores, etc., to introduce a certain play on the page along with the stage – something we find missing in the existing PaR scholarship.

In the section 'First the Senses,' we dwell on our early methodologies focused on embodying research questions and sensory data collection. Through collaborative sourcing of deep listening and daily photography exercises, we began by emplacing the body in our immediate surroundings: neighbourhoods and public spaces. Beginning with the city of New Delhi and its monuments – spaces of cultural significance – we meander through the neighbourhood, a make-shift stage inside an auditorium, and eventually chart a journey into the domestic kitchen as our final performance space in the section 'Setting the Stage'. We envisioned the performance space in line with Badal Sircar's 'third theatre' (Sircar & Chaudhuri 2021: 287–99) which suggests that the conventional, commercial and traditional theatre spaces like the proscenium (brought to the Indian subcontinent via British colonial influence) need to be reconsidered in favour of spaces that are non-commercial, intimate, minimalist, social and community-based. In addition to the usual production challenges in staging a co-devised performance (including searching for performers and artists willing to advise or collaborate on the piece, and locating a suitable venue, which we address in a later section 'Third Space Kitchen'), we wondered about how to frame the project's boundaries between the main rehearsal space of Pondicherry (with Adishakti Theatre Arts – a Laboratory for Theatre); and the eventual kitchen performance space (in New Delhi).

Adishakti had agreed to host us for a residential rehearsal process at their campus on the outskirts of Auroville – an arrangement brokered by Anusha, who had worked with them in the past. Beyond the use of an auditorium stage for rehearsal, resident actors and artists also generously agreed to attend a work-in-progress sharing to give us feedback, which we used to shape the final piece upon our return to New Delhi. Working across two cities met our need to be exposed to intracultural variety that might help us define the worlds of Insider and Outsider, even as we began to wrestle with ethical stances in inter/intracultural adaptation (Holledge and Tompkins 2000: 87–150). Each locale's selected site-specificity offered a rich and distinct art palimpsest with which we could potentially 'hybridize' our adaptation. Anusha highlighted the difference of the female experience in each location, spoken language (and resulting music and film culture), and the mode of dance/movement as aesthetic/cultural expression. We considered ideas on the then-recent events in India[4] that could provide gravity, contextual interest and relevance in the interest of increasing points of hyphenation-connection possibilities to an Indian audience.

Our early concept of collaging borrowed prose, songs from popular culture, and dance as metaphors of self-transformation addressed a particular – and exciting – challenge in our performance adaptation: how do we begin to put together culturally relevant associative elements from personal mythology to evoke multiple perceptions, emotive memories, possibilities of realities, actual sensations/reactions and personal interpretation from our Indian audience? This endeavour could be seen as a Bharucha-esque experiment, in the lineage of Rustom Bharucha's collaboration with Manuel Lutgenhorst on a tour of *Request Concert* (Bharucha 1988: 26–38), a wordless monodramatic play written by Franz Xaver Kroetz. Bharucha wrote three articles in 1990, exploring the modes of cultural adaptation used, as he outlines the process of the performances that toured in Calcutta, Madras and Bombay (*c.* 1987), each with its own unique adaptation (Bharucha 2003: 91–166). By a stroke of pure luck, Bharucha happened to be staying on the Adishakti campus when we were there, and offered his detailed feedback and response on our work-in-progress sharing, which provided a pivotal and defining intervention for the performance score of 'Measures' referred to later as the 'boiling point' moment.

> *My reflections, analysis, and words about the worlds depicted in 'Measures' – performed almost a decade ago amid a period of personal and professional upheaval – cuts through and across time and space, landing in the generosity of a wide margin. My thoughts and attitudes on the research methodology and impetus for the performance have evolved since. For instance: 'pluriverse' has replaced 'palimpsest' (a term used initially while writing about 'Measures') as I discover how worlds become un/re-made, and often not in the same space where the previous wor(l)d was formed.*
>
> *My words might find mutual collision with those of Anusha's – who proposed this written collaboration and has tasked herself with the responsibility of harmonising my contribution of marginalia into the chords of academia – a place from which I sought a maroon during my doctoral research. (Harney and Moten 2013: 22–43)*[5]

Within the methodological framework of PaR, we engage in a retelling of a process that began in 2015 and found fruition at the beginning of 2016 with a public performance. Like all collaborative ventures, it did not come without its challenges, which we have also recognized as useful to performance-making.

First the Senses – Deep Listening, Sensory Photography and Communal Meals

Both our artistic practices centre on a movement vocabulary. While Anusha's relationship with movement (explored in detail in the section 'Moving with Discomfort') comes from years of training in a dance form, Carmen's movement practice emerges from embodied choreographies involving food. While our individual movement practices are distinct, 'Measures' was nonetheless conceived as a piece that would be movement-based. Our first step in the performance-making process was to locate and reify a movement language for ourselves that we both created and resonated with. As part of our co-devising process, we both wanted to experiment with exercises that would help us define a movement process rooted in our sensorium while drawing from our memories, dreams and imaginations. We thought it might build a vocabulary that would help us ask/answer the questions around belonging to place/space and how this could be experienced corporeally for ourselves (and for the audience, somatically and sensorially). We wanted to focus on what we could hear, see, smell and taste immediately to ground our bodies in the present and move forward from there. Creating a process around the sensorium began in New Delhi with exercises drawn from practising artist-friends who responded to Carmen's request for 'a thoughtful provocation or exercise stemming from your discipline, practice, or philosophy that I/we could explore, undertake, or perform as I/we embark on this new subject/practice in new places, and in a new circumstance (of rebuilding) in my life' (email from Carmen to friendly art colleagues on 19 November 2015).

We selected three exercises that we would do together over the course of the first seven days we spent together in New Delhi: deep listening exercises offered by deep listening scholar and artist Sharon Renee Stewart[6] (after the practice developed by Pauline Oliveros and Heloise Gold), and the 'Sniff Out a Picture' sensory photography exercise from photographer Bill Crandall.[7] The third exercise emerged from our communal meals and the observational cook-alongs with our hosts, where we focused on our olfactory and taste senses, linking them to particular memories and other cross-sensory associations, notating them, and sharing these as part of our reflective exchanges. We approached these exercises of deep listening, sensory photography and communal meals as sensory data collection.

For the deep listening exercises, we began recording ambient sounds of the city while visiting various cultural sites, trying to gather snippets of the soundscape of the city that we navigate through daily and rarely, if ever, stop to keenly pay attention to. The intention was to pause and pay focused attention to something seemingly ubiquitous but easily ignored – perhaps like the labour of women in home kitchens. For instance, on our visit to Humayun's Tomb, we each took turns to sit down or stand still at various spots at the monument, claiming our spaces, to the intrigue and sometimes amusement of the other visitors, closed our eyes and listened deeply for extended periods of time, trying to notice the sounds that surrounded us. While we did feel scrutinized and also mildly ridiculed, the act of claiming a public space through being physically present, with the body that was performing what passers-by perceived as odd acts, felt, to a certain extent, defiant and liberating. To our surprise, none of the fellow visitors interrupted us by striking up a conversation (perhaps owing to our closed eyes), and none of the staff, tour guides, or security personnel questioned or bothered us during our multiple deep listening sessions – something we went in expecting. Once the initial awkwardness was overcome, it was compelling to note how a small exercise like this brought to the fore sounds of the city we were technically listening to but unaware of – the hum of the traffic, breeze, rustling of leaves, laughter, the sound of footwear dragging on the floor, metal clanks from the repair work performed by labour, glass bangles – the more we listened, the longer our lists got. At this point, it was clear to us that our piece would be wordless. If the audience was unable to see a certain action on stage (in this case, a kitchen), we would be eager for them to experience that action through the sounds it creates. We also noticed that listening to the recordings (of the exercises) later made us focus on different sounds compared to when we were listening live and in situ. We were experiencing a different or altered world through the recordings. This prompted us to imagine a simultaneous video recording being projected outside of the kitchen as we performed live within the kitchen. What would the audiences watching the live stream experience as opposed to those watching the live bodies in action inside the kitchen? In this way, the performance itself would not be the end or culmination of our research but an integral part of it. Once we left New Delhi, we would repeat the deep listening exercises in Kochi at the Ernakulam train station and in Veli near Fort Kochi at the public laundry (Dhobi Khana) – a heritage site.

Figure 6.1 Deep listening exercise at Humayun's Tomb.

Photograph by Anusha Ravishankar (2015).

Memory of a Score, Performed

> *Standing with my eyes closed on the steps of the Humayun Tomb site, I feel a tremulous breeze on my cheek and bare skin. I hear the faint pulse within my breathing—this might be my first earnest practice of meditation, which I was told could rewire the thoughts that toggled within.*
>
> *My ears catch the tinkling of a woman's laughter and other distant snatches of lively conversation. I try to listen more deeply, my knee tensing as it connects to my ankles and feet, grounded heavily on the step. I catch myself thinking about whether I'm imagining the sensations I'm experiencing in this posture. I doubt if I can really listen deeply into the architecture of this place for the dead, hear the river's ever-shifting course, or the prayers for safe passage into paradise.*
>
> *Breathing in, I can barely decipher my insides. Breathing out, I attempt to loosen and gravitate to the hum of the world outside and beyond.*

For the sensory photography exercise, the impulse to photograph texture was our means to express tactility and multi-sensorial qualities through images, and perhaps a method to then translate these into our respective movement narratives. We were most guided by the prompt at the end of photographer Bill Crandall's provocation: to receive and not just take a photo. We decided to share the photographs we took daily with each other, photographing daily for a week – pausing to receive a few photos when the alarm we had pre-set for each other would go off. At the end of each day, we would share our photographs and reflections with one another. In our early conceptualizing of what might happen in our interaction with food, kitchen equipment and utensils, we had imagined we would begin with a certain texture or quality (of food, energy or sound) that would transform through the duration of the piece. We were concerned with being able to convey to the audience the experience of sensing through touch the performance materials or props. We were aware that they would be experiencing the sounds and smells, but how could we also share with them the experience of holding and working with food apart from offering it to them to consume? Sarah Pink's work on images and the sensorium suggests that photographs are integral to multisensory environments, experienced through the interplay of the senses (Pink 2011: 4–13). They are created through movement, not as static objects, but as representations of the environments they originate from. Pink suggests that our engagement with images is both corporeal and sensory. Images, in both their creation and consumption, are therefore closely tied to the dynamic trajectories of perceiving bodies in constant interaction with their environment. They emerge from, and contribute to, the ongoing production of a particular place. Towards the end of the week-long exercise, we began to imagine the visual as emplaced and multisensorial. Photographing daily meant we had to engage with our neighbourhood of Vasant Kunj in New Delhi (where we were staying) in search of moments that conveyed touch, sound and smell. The photographs conveyed to us a quintessential south Delhi neighbourhood. From the barks of old trees that line parks of the RWA (Resident Welfare Association) colony, the steam emerging from heating the *chhole* and *kulchas* on the cart outside the gate, the sparks from metal work that was commonplace in Kishangarh village, etc.

Our third exercise of cooking and eating together began in New Delhi and continued throughout the duration of the creative and rehearsal process. This exercise was further enriched by adding collaborators who welcomed us into their home kitchens and participated in cook-alongs. We found ourselves serendipitously scheduling a cook-along with a schoolmate's mother, also living in New Delhi, who welcomed us into her kitchen and showed us how to make *besan ladoos*. Later, we also made arrangements for a cook-along with Anusha's parents during our mid-way stop in Pune, en route to Adishakti. Our observations during these sessions of kitchen-based visual and gestural data-collection introduced qualities that made their way into the performance, explored further in the section 'Becoming the Insider/Outsider'. The hollow reverberation of dabbas being filled, decanted from, or moved about inspired our choice of rhythmic drumming sounds during rehearsals, and the use of brass and other metal utensils during cooking added a macro sound structure to moments within the performance.

Carmen's participatory works, centring on food, its symbols and mnemonic links, and commensality, were created in the early days with a strong amount of stagecraft that later began to soften into intimate, social art settings resembling workshops where participants (some of whom could have wandered in from the public) were invited to take part in playful ways to feed, taste, spit and eat. Central to this praxis is a 'go-along' methodology (articulated by Margarethe Kusenbach 2003: 455–85) which she refined for her practice as a 'cook-along' to bring about embodied co-learning through following another person's cooking habitus, and hopefully, assisting in the completion of a cooking task, and perhaps a dish. 'Doing-cooking', a term by Luce Giard, in this way began to describe a keen attention to the everyday, seemingly insignificant work of cooking (Giard et al. 1998: 149). This methodology helped her to draw out and reflect stories of belonging by people more versed in acculturation through their food-making choreographies. Additionally in Lucy Long's 'Listening to the Food Voice', she demonstrates with conviction that culinary knowledge, techniques and memories can be highly relevant research data (Long 2004: 118–22).

> *Tactility, synesthesia, sentience, participation, agency, diaspora, sensoria: these residual concepts can be found within the dozen or so 'gastro-performances' that I created between 2010–2016, scattered across Washington, DC, Helsinki, Loviisa, Belgrade, Copenhagen, and Detroit.*
>
> *Making art and curating events and performances out of (dispersed) everyday food-making is my way of locating surrogate belongings and (re)directing my itinerant selves to a 'home where there are no houses'. (Le Guin, 1985).*

Moving with Discomfort

A movement-based performance aesthetic while being prescribed and taught, is one that is enmeshed in the personal and social mores (Dyer 2009: 108–23) of the body and its intentional investigation can lead the mover to explore novel ways of communicating through the body. We were aware that the creation of a 'hybrid' movement language for 'Measures' would be a process we undertook together/apart – a way to further describe the specificity of the different worlds that shaped Insider and Outsider.

> *I saw 'Measures' as an opportunity to invent a movement language that I finally had some agency in creating – a language that possibly felt comfortable and safe, a language that my body felt at ease speaking in.*
>
> *From the ages of seven to sixteen, I studied the dance form Bharatanatyam in the Kalakshetra style from a teacher who was very strict about the style being followed rigidly as she was trained by its founder Rukmini Devi Arundale. This legacy meant a lot to my teacher. As early as age twelve/thirteen, I found myself unable to identify or empathise with the themes of the stories I portrayed as a dancer on stage. For instance, the first time a dancer uses abhinaya or expressions to tell a story/perform a narrative in a Bharatanatyam recital/margam is for the choreographed piece shabdam. As a dancer, this was my introduction to acting.*
>
> *In this choreographed piece, often described by my teacher as 'light, joyful, and playful,' the principal character Krishna watches the young girls of the village disrobe and enter the river for a bath. He follows them without being noticed, and steals the clothes they've left behind on the bank of the river. The accompanying song composed by the Thanjavur Quartet (O'Shea 2007: 26–9) remains unchanged since the mid-nineteenth century. At the end of each stanza of the song, Krishna is asked if his behaviour is right or fair. The characters of Krishna and the unnamed girls from the village are all supposed to be teenagers in this episode. As a teenager myself back then, I found nothing about this character's behaviour right or fair but more importantly I did not think what the episode depicted was 'light, joyful or playful'.*

Disagreement with a teacher or guru in my class was not only discouraged, it marked your failure as a student. In any case I did not possess the vocabulary to voice my discomfort at essaying the roles of both the perpetrator and victims of what was clearly sexual harassment. I was aware that this piece made me instinctively feel uncomfortable and a little angry. I was also unable to embody a kind of mischievous play that the piece required and have fun while performing it that others in my repertory could quite effortlessly manage. This discomfort would invariably visibly translate on stage. Audiences would mention that I was not expressive, fluid, and free enough as a dancer. I contended that I must be a bad dancer and an even worse actor. Three years later, at the age of sixteen, I gave up training.

However, the years of rigorous dance training instilled in me, and it remains to this day, the need to have physical practice be a part of everyday life. Movement became the way in which I understood the limits of my body and the ways in which to use the body as a storytelling device. I don't carry anything beyond this from the training with me. Or rather I have distanced myself from the rest. This distancing became central to the process of finding a movement vocabulary for the character of the 'Insider'.

I knew Carmen would be spending time in India to rehearse, stage and perform 'Measures' with me and in preparation for this process, I returned to training in Bharatnatyam in New Delhi after a decade-long hiatus. My intention with resuming training was to acclimatise the body to the rhythm, structure and prose of the adavus[8], and to perform on stage in front of a live audience (which would include Carmen). I had found a teacher who was a recent graduate from Kalakshetra Foundation's dance school in Chennai so that I wouldn't have to learn an entirely new and unfamiliar style. I was hoping that with a young teacher I would be able to explore my previous discomforts with the stories we were essaying and the rigidity of the form itself. I was expecting that someone closer in age to me as a teacher would be able to encourage or support a conversation in the class about the ways in which our contemporary contexts clashed with the textual contexts in performance. This was sadly not the case. Very early on, I saw in my new class how certain femininities and hence certain bodies were praised and preferred over others. I realised this was systemic to the pedagogy and hence the dance form itself.

Nrithya Pillai, a practising hereditary dancer, claims that all Bharatanatyam is a mimesis of the dancing Bahujan woman (Pillai 2022). Pillai has been vocal about the invisibilisation and blatant appropriation of the hereditary dancers and especially the women dancers. Pillai's analysis of the revisionist cultural project of Bharatanatyam beginning in colonial India and finding it fruition in 1947 with the Madras Devadasi (Prevention of Dedication) Act is rooted in viewing Bharatanatyam as a 'vehicle for theatrical representations of Hindu mythology, Brahminic ethics and supremacy, and thus for majoritarian cultural politics.' Pillai's suggestion to someone like me is to just 'pause dancing'. My experience of learning and performing the form, first as a child and later as an adult, from my mixed-caste mixed-religious cis-woman identity has been in agreement with what Pillai describes not only observable through scholarship that I have encountered (see Soneji 2008: 31–70 and 2019: 112–60) but more importantly as an embodied understanding. I am very aware that I am the wrong vehicle for this form and that is communicated to me in multiple ways further alienating me from the possibility of any intimacy or relationship with this physical practice. For instance, I am told (by teachers) that I do not smile enough, I take large and loud steps while walking, surely Radha (an important character) is more 'demure', my strength training makes me more muscular, and I should look more 'delicate'. Carmen's witnessing of my Bharatanatyam performance in New Delhi as my collaborator helped me provide her with the movement context from which I planned to begin (but never return to).

Memory of Performing the Performer

Re-membering our process for rehearsing 'Measures' is an exercise in invoking the uncanny and required a surrogate: a person (me) performing the performer – with a chapeau to Marvin Carlson's definition of theatre 'as a site of the continuing reinforcement of memory by surrogation' (Carlson 2006: 2).

My mind-body needed to be goaded and bullied into performing – to orchestrate affective intensities required by the piece. In the early days of rehearsals in Kochi, I fought against the tight-lipped limpness of my imitative gestures, that I would try to train with focus and intention. None of this felt natural to me: playing a role to be seen by an audience (I much preferred being behind the scenes as a director) and the uncanny act of performing an abstracted, heightened version of my own stranger-ness and displacement. There was resistance to 'buying' my own performance quality and doubt had to be fought off with even more practice. I learnt to portray placelessness by deliberate and meditative emplacements of my body to surfaces: ground, wall, counter, sink, fridge, doorframe: 'the arch of my feet mountains, the path of my fingertips my maps,

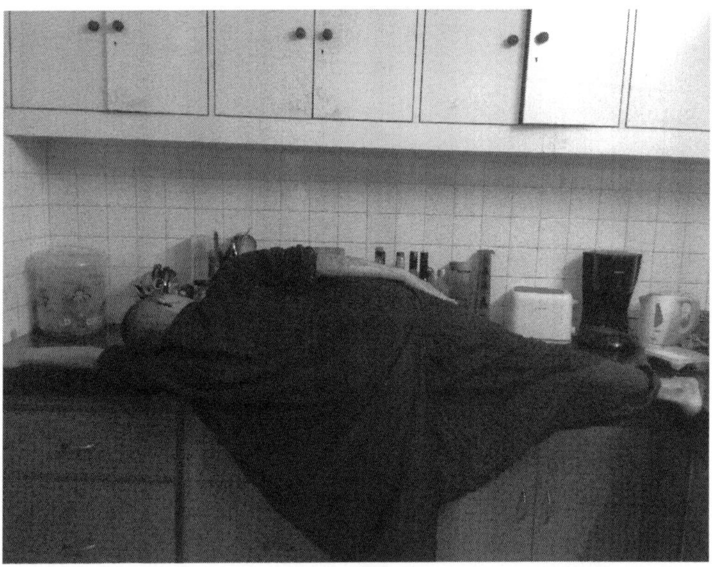

Figure 6.2 Movement rehearsals at the performance venue of the home kitchen (New Delhi).

Photograph by Anusha Ravishankar (2016).

the lines of my palms the way I go, my inbreaths and outbreaths the climate of the place I have landed' (post-embodying Le Guin's Initiation Song).

This embodied deep listening became the distance between my performance during the rehearsals at Adishakti and in the actual performance kitchen venue in New Delhi – a way to unlearn the strangeness of performance whilst at the same time unlearning to unlearn strangeness, 'as confinement becomes a way of assuming anew the predicament of ddeterritorialization: it is both I and It that travel; the home is here, there, wherever one is led to in one's movement.' (Trinh 2010: 30)

Setting the Stage

We had a stop in Kochi before arriving at Adishakti, and during this time of acclimatization, we began the process of locating the gestures of the piece. These gestures would build the score and form the base of the choreography. We would imagine ourselves in our respective home kitchens and mime all the actions we would perform to make a single meal. It was understood that the gestures were to be as close to reality as possible and not stylised in any way. This felt challenging for Anusha, as a dancer, because gestures in Bharatanatyam are highly stylized. We also video-recorded this rehearsal process to keep track of time and get a sense of the durational nature of our gestures. Finding the most compelling gestures and those that were repetitive helped us chart a rhythmic score to begin with. During this process, it became clear to us that all our gestures could be housed within the domestic kitchen, and it should be our eventual performance venue. We did not need to construct a kitchen on a stage; we would perform the piece inside an actual domestic kitchen of a home. We were also able to narrow down the utensils that we found ourselves most drawn to in our respective kitchens.

For me, gestures of using the coconut grater or 'thengai thuruvi' in Tamil, became the most recurring. I was always assigned the task of grating fresh coconut at home. To this day, when I visit my parents, I am the designated 'coconut grater' of the house. This traditional coconut grater, commonly used in South India, is an essential kitchen device used for scraping fresh coconut flesh from the shell. The grater typically consists of a sturdy, rectangular wooden base which is commonly made of hardwood. The sharp, serrated metal blade or grating surface is fixed and mounted vertically towards one end of the base. The blade is typically made of steel, with grooves or teeth to finely grate the coconut. It works with the user holding half the coconut that has been broken open against the sharp metal grating surface. The coconut is rotated or scraped over the serrated blade, producing fine coconut shreds. The process requires some strength and manual effort, and it often takes a few minutes to grate a whole coconut. I found the practice of grating a coconut with the repetitive circular motions meditative and soothing. Sometimes, I would need to pause the motion to rest my forearms or shoulders which would mildly hurt from the labour. Perhaps this was the movement language we would employ – the choreography of the everyday – women were dancing in kitchens if we would only care to look closely.

Memory of a Score from the Future

Open a cabinet to search for a plate to hold a dish of food you will make to feed your many hungry ghosts, who have followed you on your travels, whom you've shed one by one off your body as you grow new cells: did you know that your skin cells are totally replaced in seven months—all the cells that once covered a version of you are now part of dust, so many microscopic deaths that un/make you.

Oh? what else have you left behind besides cells, hair, dust, books, tools, gifts, promises, spiteful words, rituals, love, homes, livelihoods, companions, commitments, memories, now tightly buried in ash.

Light a stove, to fry an egg, a piece of toast, some bits of onion, a sliver of ginger, garlic if you have some, why not a bit of chilli for the smell of something familiar – food as incense to these forgotten ghostly selves.

Little do you know, as you fish a fork from the drawer, that in the coming years and in the many kitchens you'll use as a stage, you'll fry noodles made of string, turn them into a soupy dish served with hand-folded dumplings stuffed with red paper pulp and fragrant spices – food to feed these ghosts more earnestly, begging them to let you rest.

By the time we arrived at the Adishakti campus on the outskirts of Auroville, we had amassed a bank of experiences, experiments and sensorily derived data to draw from during this period of preliminary work. The campus, sprawling across three acres, was the ideal space for artists looking to create work. It was equipped with an amphitheatre, a fully equipped auditorium – Koothu Kovil, access to their archives of rare performance recordings, a library of theatre and performing arts publications, an opportunity to collaborate with other resident artists who were training in a host of creative and physical practices, and many of whom trained at the neighbouring *kalaripayattu*[9] ashram, which is where we also began to train to give structure to our days and our physicality a bit more strength. The two weeks spent at Adishakti were immersive and rigorous. After our morning *kalari* practice, we spent hours on the stage blocking and rehearsing our piece, and in the time away from the stage, we were swimming, stretching, reading, writing and engaging with other artists and scholars, sometimes joining them in their practices of Carnatic singing, yoga or drumming. Some of the artists on campus generously volunteered to be part of our rehearsal process. Visiting actor Steina Öhman offered to play the *mizhavu* – a metal pot-shaped

Figure 6.3 Rehearsals at the Adishakti Laboratory for Theatre Art Research. Photograph by Anusha Ravishankar (2015).

percussion instrument from Kerala – during our rehearsals to provide a beat or rhythm to our movements.

By the end of the first week, we had a narrative shaping up, a sense of the movement's impetus and an abstract background for each of our characters. We used the kitchen utensils from the campus kitchen and were able to create a kitchen on stage by marking its borders with stools available in the green rooms.

Becoming the Insider/Outsider

What would occur in the kitchen we were planning to inhabit? What would the actual kitchen look like? And who were the two characters housed within this kitchen? This was, personally, the most difficult part of building this piece. Developing the character of 'Insider' meant having to confront and embody attributes and emotions that were challenging to sit with.

For instance, I was very aware of how Carmen was being perceived as an Asian woman with a shaved head in the village of Edayanchavadi in Villupuram district, which was near the Adishakti campus and frequented by us during our stay. I knew the stares she was subjected to went beyond the initial fascination. I wanted to investigate the possibilities of various meanings her body brought to the spaces she inhabited. The actor portraying the 'Insider' had to not only hold space for possible prejudices but also needed to be able to perform them. I found myself nervous about this portrayal, in addition to the discomfort of identifying these prejudices within myself in order to bring truth to my performance.

Working with any collaborator this intimately every day for over a month might invite moments of friction. We were not immune to those, in spite of our friendship – and it was complicated by the fact that Anusha felt beholden to Carmen's safety as a first-time visitor to New Delhi, as her host. A Thomson Reuters Foundation global survey of experts in 2018 found India to be 'the world's most dangerous country for women due to the high risk of sexual violence and being forced into slave labour'. We had an open conversation about it, and Anusha was able to relax from imposing her fears during our travels to the south of India. During our stay in Kochi, we would set out at the start of the day independently, making our way through Fort Kochi, and regroup for our rehearsals and for the meals we shared together. We watched performances together. Sharing rooms and homes during the entire creation and rehearsal process also meant that we were constantly in close quarters, having to navigate an intimacy not hitherto experienced with each other. While this had its own challenges, we agreed that it felt like it enriched our dynamic as performers.

I was also keenly observing the participants in Carmen's kitchen cooking sessions – noting their reaction to having her in their kitchen spaces. It helped greatly that I was intimately aware of how my mother feels about having a guest in her kitchen space even if she did not communicate her feelings through words or gestures. For instance, I was aware of how particular she was about how her glass jars of ground spices were arranged. It might seem like a random assortment, but I know how she has committed to memory which spice is on which shelf and in what order. If that order is disrupted, my mother can go very quickly from mild irritation to deep disappointment and eventually demanding that you leave her kitchen quite sternly. It was the same for how her kitchen utensils and equipment were arranged. I must admit that when I sensed that Carmen could possibly disrupt the order, I said nothing, almost hoping to witness how my mother would respond. Carmen was treated with a lot more generosity. I wanted to embody these spoken and unspoken rules that women have about their kitchen as the 'Insider'.

Most upper-caste Hindu kitchens in Indian apartments also have small home shrines with idols, which are often offered sweets or other treats during the morning and evening prayers. The preparation and offering of these follow certain rituals, which Anusha wanted to bring to the kitchen of the 'Insider'. However, we felt this could be achieved symbolically. Stepping into the final performance space – which was the home kitchen of a theatre artist

Figure 6.4 Constructed shrine.

Photograph by Anusha Ravishankar (2015).

and close friend in New Delhi – we knew a shrine would have to be built because she didn't already have one. Anusha emptied a kitchen cabinet, placed a stone in it to signify the idol, and arranged some fresh flowers around it.

We remember Anusha being surprised by feeling protective about this shrine when someone from the house (whose kitchen it actually was) decided to rearrange it or move something in it. We had cooked in that kitchen before. Anusha, in particular, began to think of it as hers, and the more she arranged the grated coconut, raw rice, flowers in the shrine and utensils, the more she found herself becoming miffed with Carmen moving things around, which the 'Outsider' would inevitably do. She did not fight that feeling, allowing it to grow and find its utility and moment in the performance.

Excerpt from 'unmade/untitled,' a performance that was birthed upon leaving New Delhi

Where I used to come from, objects seemed to rust easily, being so close to the equator—the humidity working to speed the work of salty corrosion. On hot days, you can almost feel salt grains on your cheeks, and bare skin.

Each time I visit, I gulp hungry deep breaths ... as if breathing in the sea was one way to remind my body to remember what it feels like to desire to be in place.

But each time I come back to this place, having roamed a little more, having swum in and inhaled the seas of other places, I have to work harder to remember.

I do this by secretly licking the salt off my skin, tasting sweat, sea, debris from all my wanderings.

Boiling Point

From the moment the piece begins, the two characters exist in a movement language of separateness. Tension builds, interactions turn into interference and each micro-gesture creeps to a boiling point.

That meeting in a violent moment – where there is a channelling of Insider's anger that is inflicted on the rag-body of Outsider – is a fissure: a release of emotional violence, an easing of tension. This moment came together as a result of creative input/push from Rustom Bharucha while at Adishakti. When we shared a work-in-progress version with him, our moment of collision was a cacophony created by utensils falling to the floor. This noise/moment breaks the separateness between Insider and Outsider (who, until this moment, go about their routines without acknowledging each other directly). The two seem only to be aware of each other's presence through the disruptions caused within the space and via utensils and food items being tampered with or misplaced. At the moment of collision, they are forced to confront each other by acknowledging the other's body in the space. Bharucha encouraged us to channel the pent-up frustration and anger present in that moment not only through the falling of utensils but through our bodies. How would Insider enact her rage upon the Outsider's body? And how would Outsider respond?

> *I wanted 'Insider' to represent the dominant narratives of a place but felt slightly ashamed about it. She was someone who was reluctant or embarrassed to admit that she was possibly xenophobic, casteist, and racist. When she couldn't mask it anymore, she revealed exactly who she was by violently showing the Outsider what she thought of her. I found that scene—the boiling point—the hardest to perform. I couldn't imagine being that angry with anyone. I also experience rage internally; I rarely, if ever, show it. This expression of rage was alien to me, perhaps why it was eventually easy to perform. I thought of it as a novel experience. To use the body of another person as a rag to clean the kitchen floor demonstrated feelings of disgust and dehumanisation. I found it difficult to channel those towards Carmen. I tried to focus on the act rather than the feeling. I wasn't able to push her body to the floor and drag it across multiple times unless I generated a force that was coercive and overpowering. I had to be violent in order to achieve the movements we had planned for the sequence. The emotions had to*

follow the movements. With every rehearsal, I found myself being able to embody, through movement and breath patterns, emotions that I could not generate independently.

In hindsight, our work explored the 'riss' between ritual and the everyday and what emerges from that split. The same kitchen that houses the Gods holds invisible walls that separate our two characters, who exist in a slice of this everyday and slip into rituals within the everyday. There is a doubling of experiences across what Edgerton and Roberts juxtapose as Bourdieu-ian 'habitus' and internalized 'cultural capital', seen here as the ability to belong versus the ability to un/learn strangeness. 'Despite the seemingly repetitive character of its theme and variations, the take of hyphenated reality continues its hybridising process. It mutates in the repercussive course of its reproduction as it multiplies and displaces itself from one context to another. It is, in other words, always transient' (Trinh 2010: 34)

Eventually, the 'boiling point' became the defining amalgamation of the new movement vocabulary we had created for 'Measures' – informed by the sensory, choreographed to resemble the everyday movement of women's bodies, propelled by the emotional truth held within the body, bringing to the fore narratives explored through breath and gesture.

Third Space Kitchen

The kitchen, in Indian households, is predominantly identified as a space that is the domain of women; it is also where they labour. The kitchen is, at the outset, subject to Lefebvre's 'comparable approach' to 'social space,' where the space isn't the objects or things contained within it but the social relationships and norms embedded within it (Lefebvre 1991: 89). The ways in which everyone else (apart from the women) relates to the kitchen in the household are transitory and utilitarian. For them, the kitchen exists on the margin. If women principally exist in this (kitchen) space, do they not then also exist in the margin? While the Outsider might believe that she is operating from the margin, trespassing into the Insider's space, the kitchen emplaces both of them on the margin of the household. We think of this margin, as bell hooks suggests, as a site not of 'deprivation' but one of 'possibility' (hooks 1989: 15–23). For hooks, the margin is a space where the oppressed can find solidarities and connections with others. By using these networks, she suggests that the margin becomes a space from which to activate modes of resistance and radical inclusion.

> As such I was not speaking of a marginality one wishes to lose –to give up or surrender as part of moving into the centre –but rather as a site one stays in, clings to even because it nourishes one's capacity to resist. It offers to one the possibility of radical perspective from which to see and create, to imagine alternatives, new worlds.

Edward Soja further imagines the distinctions between these spatial binaries (centre and margin) as dissolving into a 'thirdspace' by describing it as a 'strategic location for exploring postmodern culture and seeking political community among all those oppressively peripheralized by their race, class, gender, erotic preference, age, nation, region, and colonial status' (Soja 1998: 137–9). Diana Fuss complicates the inside/outside binary effectively by suggesting that perhaps, as in the kitchen, it must be negotiated with in order to confront it. Fuss finds that the binary is impossible to dismantle entirely but can be expanded to reveal its multiple meanings and implications.

> The figure inside/outside cannot be easily or ever finally dispensed with; it can only be worked on and worked over-itself turned inside out to expose its critical operations and interior machinery ... The problem, of course, with the inside/outside rhetoric, if it remains un-deconstructed, is that such polemics disguise the fact that most of us are both inside and outside at the same time. (Fuss 2013: 1)

The repeated assembly, disassembly and interplay of texts, materialities of food, bodies and utensils bring and attach significance and meaning to the space and time of the kitchen – a constellation of relationships in place beyond the exact location itself, where we gain more understanding from how these relationships weave and intersect (Massey 2005: 9–13). The dynamics and sensorium of the kitchen become a 'worlding' for the strangeness, belongings, implicit disharmonies and imperfect understanding embodied by Insider and Outsider, for 'feelings and tastes are an activity of "worlding" that renders life as this life and not another, and renders time as this time and not another' (Highmore 2016: 146).

While we knew from the very beginning that we would be performing the piece in a kitchen, we were unprepared for how many variables come with performing in an alternative performance space – the actual home kitchen. We had interruptions sometimes by the homeowners, their dog, the cook, etc., and we had to stop our rehearsals to accommodate them. We were technically in their space. It didn't matter how well we tried to plan our rehearsal times. Because of how the kitchen was built, the audiences would be peeping in from the entrance, which was large enough to accommodate the body of one adult passing through at a time. It felt like the audiences were voyeurs and privy to something intimate and personal. We did not have any lights apart from the ones installed and used by the residents for everyday use. We invited a small group of fewer than fifteen people to watch our performance. Some were educators, students, people working in public policy, theatre makers and dancers. We encouraged them to share feedback. Incidentally, the people more concerned with the lack of theatre lights and sound systems ended up being the theatre practitioners. The dancers thought the movements were not stylized enough. It was the educators, students and policymakers who spoke to us about what they saw happening between the two bodies. They drew connections to themes of colonization, xenophobia, racism, violence and domestic labour. The kitchen becomes a metaphor for the homes, cultures and national identities inhabited by Insider and Outsider. This work employed different movement styles, a minimalist soundscape, material language and the sensory experience of the everyday kitchen to depict the uneasy intersection between the two characters. Insider's distracted, exhausted, repetitive actions of absent care and attachments to ritual are seen (by the Outsider) as curious, who is unintentionally disruptive of these systems of worship, order and work. Outsider's actions, too, are an attempt at cautious home-building, deployments actively fostered to confront a new place and a new time, and to try and secure oneself a homely life within them (Ghassan Hage 1996: 463–85).

For revisiting and the retelling of 'Measures', we collage our writing with nonlinear memories and fragmented narratives, constructed by reflecting on an archive, recollections of rehearsals and conversations. In consolidating the various elements of our respective archives into processes of sensory data collection, embodied practice as research, and collaborative performance-making, we have been able to articulate for ourselves the journey of a performance piece close to our hearts. We have put the elements together (as very different individuals today) from those that inhabited the bodies of Insider and Outsider – a coming together of familiar strangers to weave a narrative of ourselves, our memories, and the places we've lived in – much like we put together our performance a decade ago.

Notes

1 A hyphenated identity we both felt comfortable inhabiting back then.

2 Hutcheon's theories of adaptation are particularly helpful to our approach because of her consideration of intertextuality within the act of adaptation. She argues that every adaptation is in a dual relationship with the source text: the adapted work is new and distinct while simultaneously refers back to and transforms the original. The process of adaptation involves reinterpretation, reconstruction and often

re-contextualization, which reshapes how the audience understands both the original/source text and the adapted version.

3 We were taught a performance practice module at the University of Warwick by Kapur as students as a part of which we created individual and collaborative performance pieces. Her contribution to shaping our research methodologies as a mentor is vital.

4 In 2013, following large-scale protests from Indian citizens the laws pertaining to rape and sexual assault were amended to increase punishment, make it easier to report instances of sexual assault and scrapping of unscientific tests to ascertain the validity of the survivor's allegation. The Criminal Law (Amendment) Act, 2013 also known as the Nirbhaya Act led to renewed conversations in professional and personal settings around women's safety in public and domestic spaces.

5 Reasons for distancing myself from academia might be understood through an engaged reading of *The Undercommons: Fugitive Planning and Black Study.*

6 'Public Spaces listening: 10-minute exercise performed at least two times per week.

Materials needed: your body and a notebook with pen, optional: recording device

* Find a spot in the public space (indoors or outdoors) where you can sit or stand quietly with your eyes shut.

* Begin with listening to your breathing and feeling your heartbeat and slowly expand your listening as far as you can. Listen in a sphere, front, back, up and down. Welcome all sounds as they enter this listening sphere and leave it. Give them your attention and allow them to come and go.

* Take another 3 minutes to do stream-of-consciousness writing/ drawing about your listening experience in your notebook.

* Development: Open your senses to listen with your ears, with your skin and with your entire body (the fluids of your body might even vibrate).

Variation: rotate 90 degrees every one minute or so, so that you listen (something like: North, South, East, West)

Variation: record this process and listen to your recording later. Notice the similarities and difference of your live listening with the listening of the recording. Are there sounds that now stand out more or are masked?'

7 'For one week, once a day, take out your phone camera and "sniff out" a picture wherever you happen to be. Don't think "what usually makes a good picture?" Look around, shut out distractions, get in tune with your intuition, and see what offers itself. No matter how humble. It can be simply light, an object, a juxtaposition, a human moment, a texture ... The goal is to follow your nose, so to speak, until you feel like you are receiving the image as opposed to constructing it. To have that little moment of "oh!", where all you need to do is press the button to channel/receive/embrace what is on offer. Ideally, in that moment you have become fully sensitized, alert, and focused, yet also passive. Again, you are receiving, not grabbing'.

8 The *adavu* is a single step, the basic unit or building block of footwork, hand gesture and basic posture for the Bharatanatyam vocabulary.

9 Indian martial art that originated in Kerala during the eleventh–twelfth century CE.

References

Allsop, Ric and André Lepecki. 2008. 'Editorial: On Choreography'. *Performance Research* 13(1): 1–4.

Bharucha, Rustom. 1988. 'Theatrework: The Request Concert Project'. *Performing Arts Journal* 11(1): 26–38.

Bharucha, Rustom. 2003. *Theatre and the World: Performance and the Politics of Culture*. London: Taylor & Francis.

Carlson, Marvin. 2006. *Speaking in Tongues: Languages at Play in the Theatre*. Ann Arbor, MI: University of Michigan Press.

De la Cadena, Marisol and Mario Blaser (editors). 2018. *A World of Many Worlds*. Durham, NC: Duke University Press.

Double Edge Theatre. 2024. 'Double Edge Theatre'. Available at: https://doubleedgetheatre.org/training/ (accessed on 2 March 2025).

Dyer, Becky. 2009. 'Merging Traditional Technique Vocabularies with Democratic Teaching Perspectives in Dance Education: A Consideration of Aesthetic Values and Their Sociopolitical Contexts'. *The Journal of Aesthetic Education* 43(4): 108–23.

Edgerton, Jason D. and Lance W. Roberts. 2014. 'Cultural Capital or Habitus? Bourdieu and Beyond in the Explanation of Enduring Educational Inequality'. *Theory and Research in Education* 12(2): 193–220.

Fuss, Diana. 2013. *Inside/Out: Lesbian Theories, Gay Theories*. New York: Routledge.

Giard, Luce, Pierre Mayol and Michel de Certeau. 1998. *The Practice of Everyday Life, Vol. 2: Living and Cooking*. Minneapolis, MN: University of Minnesota Press.

Hage, Ghassan. 1996. 'The Spatial Imaginary of National Practices: Dwelling – Domesticating /Being – Exterminating'. *Environment and Planning D: Society and Space* 14(4): 463–85.

Harney, Stefano and Fred Moten. 2013. *The Undercommons: Fugitive Planning and Black Study*. Wivenhoe: Minor Compositions.

Highmore, Ben. 2016. 'Formations of Feelings, Constellations of Things'. *Cultural Studies Review* 22(1): 144–67.

Holledge, Julie and Joanne Tompkins. 2000. *Women's Intercultural Performance*. London: Routledge.

hooks, bell. 1989. 'Choosing the Margin as a Space of Radical Openness'. *Framework: The Journal of Cinema and Media* 36: 15–23.

Hutcheon, Linda. 2006. *A Theory of Adaptation*. New York: Routledge.

Kroetz, Franz Xavier. 1987. 'Request Concert'. Directed by Rustom Bharucha and Manuel Lutgenhorst. Performance in Calcutta, Delhi, and Bombay.

Kusenbach, Margarethe. 2003. 'Street Phenomenology: The Go-Along as Ethnographic Research Tool'. *Ethnography* 4(3): 455–85.

Le Guin, Ursula K. 1985. *Always Coming Home*. New York: Harper & Row.

Lefebvre, Henri. 1991. *The Production of Space*. Oxford: Blackwell.

Long, Lucy M. 2004. 'Learning to Listen to the Food Voice'. *Food, Culture & Society* 7(1): 118–22.

Massey, Doreen. 2005. *For Space*. London: SAGE Publications.

Oliveros, Pauline. 2005. *Deep Listening: A Composer's Sound Practice*. Lincoln: iUniverse.

O'Shea, Janet. 2007. *At Home in the World: Bharata Natyam on the Global Stage*. Middletown, CT: Wesleyan University Press.

Parasecoli, Fabio. 2019. 'Eating Power: Food, Culture, and Politics'. In *Postdisciplinary Studies in Discourse*, edited by T. Marttila, 129–53, Discourse, Culture and Organization. Cham: Palgrave Macmillan.

Peterson, Indira Viswanathan and Davesh Soneji. 2008. *Performing Pasts*. New Delhi: Oxford University Press.

Pillai, Nrithya. 2022. 'Re-casting the Narrative of Bharatanatyam'. *Economic and Political Weekly* 57(9). https://www.epw.in/engage/article/re-casteing-narrative-bharatanatyam.

Pink, Sarah. 2011. 'Sensory Digital Photography: Re-thinking "Moving" and the Image'. *Visual Studies* 26(1): 4–13.

Sircar, Badal and Paromita Chaudhuri. 2021. 'The Language of Theatre'. In *Critical Discourse in Bangla*, edited by Subha Chakraborty and Subrata Sinha, 287–99. India: Routledge.

Soja, Edward W. 1998. 'Thirdspace: Journeys to Los Angeles and Other Real-and-Imagined Places'. *Capital & Class* 22(1): 137–9.

Soneji, Davesh. 2010. 'Śiva's Courtesans: Religion, Rhetoric, and Self-Representation in Early Twentieth-Century Writing by Devadāsīs'. *International Journal of Hindu Studies* 14(1): 31–70.

Soneji, Davesh. 2019. *Unfinished Gestures: Devadasis, Memory, and Modernity in South India*. Chicago, IL: University of Chicago Press.

Trinh, Minh-Ha T. 2010. *Elsewhere, within Here: Immigration, Refugeeism and the Boundary Event*. London: Routledge.

7

Narratives and Reflections of Mothers in Indian STEM Academia

Sre Ratha

Introduction

Female representation in STEM (Science, Technology, Engineering, and Mathematics) is consistently increasing in India. In comparison to developed countries, women's enrolment rates in STEM education are much higher in India. As per data from a recent World Bank report, 43 per cent of total graduates in STEM in India are women as opposed to 34 per cent in the United States, 31 per cent in Canada and 38 per cent in the United Kingdom. Despite such high representation of women at the graduate level, these numbers do not translate with respect to representation in research labs and academic institutes. Demonstrating the gender gap in in STEM research and academia, the World Bank report also points out that only 14 per cent of India's women hold positions of scientists, engineers and technologists at research institutions and universities. Though the Government of India through the launch of several schemes is constantly encouraging more women to pursue research careers in STEM, nonetheless, as the data shows women representation continues to remain low in STEM research establishments and higher education organizations. According to a recent media report, India's premier technological teaching and research institutes have very low percentages of women professors and none of these top-tier science and technological institutes have ever appointed a woman director

(Sahoo 2021). Likewise, Kaur and Kapoor (2022), in their recent paper, argue that women's underrepresentation in leadership roles in STEM disciplines could be attributed to culturally rooted traditional gender roles and the pressure of marriage and motherhood on Indian women. Highlighting the impact of motherhood, other studies across the globe have also pointed out that working mothers are the most affected cohort in STEM fields (Thébaud and Taylor 2021: 397–400; Doerr 2022: 713–20). Having to contend with workplace expectations as well as being primary caregivers for their children, working mothers in STEM organizations are routinely passed over for promotions (Fuller 2018: 1446–73; Beutel and Scheiffer 2021: 795–811). In a shocking incident in one of the leading scientific institutes of India the tenure application of a female faculty was rejected for not having published during her maternity leave (The Wire 2018). Commenting on the normative expectation of motherhood, media reports from India suggest that women often have to give up their careers and drop out from PhD programmes and faculty positions as that coincides with their child-bearing age (Masoodi 2016; The Wire 2018). The crucial point to understand is that even though women are opting to pursue STEM disciplines at the higher education level, nonetheless, they opt to exit STEM careers in India owing to the lack of permanent jobs and not receiving their deserved promotion. This problem is particularly acute for mothers as STEM disciplines require long hours devoted to innovation and they are often expected to choose between their careers and motherhood duties. Despite the impact of motherhood on a woman's career choices, the intersections between motherhood and STEM academia have remained under-examined in Indian scholarship. Taking note of this research lacuna, this chapter through qualitative interviews intends to examine the myriad challenges faced by working mothers in STEM academic institutes. Additionally, this chapter also aims to understand the various strategies adopted by these mothers to sustain themselves in this demanding field of research. Using a feminist science studies approach, this chapter will examine how motherhood is a key factor in shaping the experience in STEM. Finally, this chapter will attempt to discuss a policy framework for working mothers opting for a career in STEM.

Investigating the link between motherhood and female labour force participation, scholars from different parts of the globe have discussed and underlined the concepts of the 'glass-ceiling effect' and the 'motherhood penalty'. By definition, the 'motherhood penalty' implies that mothers are discriminated against and receive lower wages in their workplaces as organizations doubt their competence and commitment (Correll et al. 2007: 1299–336; Budig et al. 2012: 166–91; Kahn et al. 2014: 60–7; Yu 2018: 1070–83; Ray and Samanta 2020: 39–49; Hsu 2021: 2–4). Likewise, the 'glass ceiling effect', entails that women are often prevented from holding positions of power in their workspace as women are expected to maintain a work-life balance (Morgan 1998: 480–90; Purcell et al. 2010: 711–13; Kaur & Kapoor 2022). In addition to these studies which have explored the 'motherhood penalty' and the 'glass-ceiling effect' in different contexts, recent studies have also highlighted the many structural barriers that mothers have to navigate in STEM organizations. For instance, in their study conducted among PhD students and Post-Doctoral scholars in natural sciences and engineering departments in the United States, Thébaud and Taylor (2021: 395–419) through their theoretical framework of 'specter of motherhood', argue that cultural biases often discourage young women of childbearing age pursuing their careers in the fields of science and engineering from having children. In particular, the authors point out that the respondents in their study believed that motherhood is to be feared as chances of mothers getting hired or producing higher standards of work were less owing to their responsibilities towards their children (Thébaud & Taylor 2021: 395–419). Likewise, in another US-based study Doerr (2022: 710–23) points out that women with PhDs in science and trained as researchers are often hired as contingent teaching faculty working on yearly contracts for a lower pay. Through the narratives of the study respondents, Doerr (2022: 710–23) argues that many of these women accepted these positions as they were married to high wage earners and because they were already mothers and had to perform their motherhood responsibilities as well. Since this position gave these women a certain amount of flexibility, they accepted this low-status position, and the university continued to benefit by paying them a relatively low salary (Doerr 2022: 710–23). Noting the dearth of doctoral students who chose to be mothers while pursuing their doctorate degree in the field of STEM, Simon and Swindle (2024: 551–69), highlight the case study of a Black new mother and her experience of being a new mother as well as her being a doctoral STEM student at the same time. Through this case study, Simon and Swindle adopt a Critical Black Feminist lens and point out that this case study warrants individual attention as finally after seven years their participant had to leave the doctoral programme without completing her degree. In particular, the authors note that their participant experienced a 'marked shift in her advisor's attitude' after her pregnancy became public knowledge. Though the participant tried her best to adhere to institutional norms, nonetheless, the dominant white supremacy in STEM institutes ultimately pushed her to leave the doctoral programme (Simon and Swindle 2024: 551–69). Apart from these studies that have highlighted the institutional and cultural challenges faced by mothers in STEM establishments, another set of studies has demonstrated how the recent pandemic, Covid-19 had a significant impact on female scientists. As the pandemic led to loss of childcare several female scientists and faculty members who had children were impacted negatively as they tried to fulfil their parenting role as well as their organizational demands (Fulweiler et al. 2021; Dunn et al. 2022: 573–9).

Drawing from these studies, which have elaborated on the inherent gender biases in STEM research organizations and universities, the present chapter seeks to understand how Indian mothers who are working in STEM academia navigate the system on a regular basis. This chapter proposes to explore the motherhood experiences of women who, on one hand, are forerunners of change by holding academic positions in the male-dominated field of STEM, while, on the other hand, they are complying with the normative role of motherhood. Relying on qualitative interviews, this chapter aims to unpack how these mothers in STEM academia strike a balance in their dual roles, which require them to challenge as well as conform to societal norms at the same time.

In addition to expanding on the scholarship on STEM mothers, this chapter also borrows from the interdisciplinary field of feminist science studies and uses this as a theoretical backdrop to explore the debates that have shaped this field. Broadly, feminist science studies have explored the relationship between race, class, gender and science. This field has been expanded significantly by several feminist scholars. Evelyn Fox Keller (1987: 37–45) noted that science is mostly defined as a masculine discipline which is dominated by men. Given her background in physics and interest in biology and working as a woman scientist she indicated that owing to societal roles of men and women which relegate men to the public sphere and women to the private sphere, women are usually identified for their reproductive roles and omitted from the discipline of science. Hence through her works she urged that feminism and science come together to raise consciousness among scientists and pave the path for a more transformative nature of the discipline of science (Keller 1982: 589–99). Donna Haraway (1987: 103–18), in her seminal work titled *A Cyborg Manifesto* pointed out that the creation of cyborgs will eliminate dichotomies such as nature/culture, mind/body, and idealism/materialism, critiquing the frameworks of traditional feminism which established that women are more invested in managing everyday life and hence occupy an epistemological privilege. Instead, she argued that with the rise of cyborgs, individuals can form groups based on their affinities and choice (Harraway 1987: 103–18). Expanding the scope of the field, Subramanium (2009: 951–77) suggests that in their analysis, feminists should not view science as a homogeneous discipline and flesh out the plural nature of science. In particular, she notes that instead of asking if feminism can change science, it is important to look into the many intersections and interdisciplinarities to refashion feminist science studies at a global level. Chadha and Achuthan (2017: 33–6) in their paper interrogate 'science' as a discipline and indicate that it is important to adopt a reflexive lens to address gender biases in STEM disciplines. Women often question their own self-worth as they enter the field of science as young researchers as upper-caste men dominate the research and teaching institutes of science in India (Chadha and Achuthan 2017: 33–6). In the Indian context, caste is a significant factor, and it plays a key role in shaping the experience of men and women in science. However, as this chapter could not look into the intersections of caste and gender and the focus was on science and motherhood, this aspect of the theoretical framework, while acknowledged, will remain under-examined in this chapter. Although this chapter does not introspect the intersections between caste, religion, motherhood and STEM, as the main focus is on the many challenges that STEM mothers experience on a regular basis. Additionally, as highlighted the aim of this chapter is to apply the feminist science studies theoretical framework to the present chapter and explore policies that will enhance the support for STEM mothers and contribute to the conversation on women in science.

The present chapter relied on qualitative methods to recruit participants, conduct interviews and analyse the transcripts as well. As a first step, the author began by writing emails to female faculty members of top public technological and research institutes in India. As the majority of the esteemed institutes are public, the chapter was limited to the public universities and institutes of India. Female faculty members who are mothers and working in private organizations were not contacted for this chapter. The email was sent to all female faculty members in all such institutes, as the author had no prior knowledge of who in their sample population would be a mother.

The inclusion criteria of the chapter detailed that respondents who were mothers with a doctorate degree, belonged to the age group of 35–45, had access to the internet, spoke English and could devote around four to five hours of their time would be interviewed. Owing to the pandemic and travel restrictions, all the studies were conducted through Google Meet and prior consent was obtained to record the interviews.

Though each of the respondents' narratives and journeys in STEM academia had their own uniqueness, nonetheless, after the transcription of the interviews, the author identified common traits in the narratives. There is not much scholarship on STEM mothers in the Indian context, and one of the main aims of this qualitative chapter is to build more literature on this research topic. In the process, this chapter particularly uses 'narratives' in qualitative methods as a lens to capture the experiences of STEM mothers in Indian academia. In most qualitative studies, scholars have used in-depth interviews as a tool to conduct research. However, in this chapter, narratives have been used to understand the main themes of the interviews, as the main goal of this chapter is to outline the everyday struggles of STEM mothers.

All the names of the respondents have been changed to maintain the anonymity of the respondents.

STEM Academia and Motherhood: Juggling Multiple Roles

One of the major points found in all the narratives was that all the respondents took around six to seven years to finish their doctoral degrees. After completing their degrees, all the respondents held post-doctoral positions for two to three years before applying for a faculty position. All the respondents mentioned that in India, completing a PhD in STEM is quite time-consuming. Additionally, it is mandatory to hold a post-doctoral position to be able to apply for a faculty position in a public organization. As a result, several of the respondents delayed their pregnancies and were ready for a child only when they were sure of their financial viability and job security. However, all the respondents emphasized that it was their decision to become mothers, and they were not really pressured to embrace motherhood due to family expectations. Hence, one of the main challenges for all the respondents was to maintain a balance between their new positions and fulfil their duties as new mothers. As in most cases, motherhood and the job position happened around the same time, the respondents found it hard to play dual roles.

> Honestly, I do not know which role to manage first. As a PhD student in STEM, I wanted to do so many things. I had many aspirations. So, when I got this job, I was very delighted. All my hard work had paid off. But then, eventually, my husband and I started talking about starting a family. I knew it would be difficult and that I would have to manage both fronts. Nonetheless, I went for a baby because I was getting older, and how much longer could I delay? I obviously did want a child. However, after becoming a mother, life has been a rollercoaster. I have huge administrative responsibilities, and it becomes extremely challenging when the child is ill. (Vaishnavi, aged 39)
>
> Administrative responsibilities are a major responsibility in academia. I realized that only after becoming a faculty member in a public organization. As most committees need women representatives and there are very few women faculty members in STEM, most of us get over-represented, and hence our workload never really ends. I really have to factor in my leaves, as most of my leaves are used for childcare. Despite having different support systems, a child needs his/her mother. As a result, I hardly pay any attention to my own health. (Juhi, aged 35)

In addition to their teaching requirements, as indicated by the respondents, they were members of several committees and were expected to perform a significant number of administrative responsibilities. As part of these responsibilities, all the respondents had to attend several meetings and prepare numerous documents. Thus, in the majority of cases, the health of the respondents took a backseat, and many of the respondents had gone to work with mild fever, cold and cough.

The other major challenge pointed out by the respondents was that, as STEM was dominated by men, they always feared that they would be chided for prioritizing their child. All the respondents pointed out that, in most cases, their older male colleagues, who were often the heads of the departments and the deans, would compare them to their spouses, who were homemakers or schoolteachers and were able to devote a significant amount of time to their kids.

> These days, most women are opting not to have children so that they do not have to slow down in their careers. Motherhood is a difficult choice. It slows you down, and you have to prioritize your child. The biggest challenge is the pressure to constantly publish. As a result, we all take around three to four PhD students. Once you take students, you have to ensure that they publish.As a mother, I think sometimes my identity as an academic is not seen enough. No matter how hard I work, I am scared to apply for a promotion because I am not published enough. None of my students have graduated, and so I am always unsure of myself. I truly think that there should be different regulations for mothers, and the period of maternity leave should not be seen as a time when we are expected to publish and do research.These days, parents do not want to live with us permanently, and we have to devise our own network settings. Hence, it is very challenging, and there are so many meetings that one is expected to attend that I wonder how I will ever make time to do undisturbed research. (Smita, aged 40)
>
> They do not say this to my face, but I know in their heads they want us to fail. They think we, as women, are not capable. Hence, I get this strong urge all the time to outperform. I feel I have had similar struggles and challenges, so why should I not get my due? I deserve more. It is not fair that I should be blamed for becoming a mother, and my professionalism should be doubted. That is what hurts me the most. I also feel there is very little understanding of the real struggles of mothers.I know, for example, one mother who

> has a PhD. She quit her job because of the lack of child-care support in the institute. As much as I know, her parents passed away in the pandemic, and it was hard for her. She did not get much support, and so, finally, she had to quit. Even today, as you know, motherhood is the real reason why women continue to leave their jobs because society expects them to be mothers first and then anything else. (Sulekha, aged 37)

The women felt that they held equal positions to men and that they had to assert their identities, as most of the top positions were held by men. All the respondents complained that the men hardly understood their problems and would always give an example of their own wives to prove that women are better off in the domestic sphere. Additionally, the respondents stated that they faced competition from other childfree and single women from their institutes, who were always available for meetings at odd hours. Though none of the women considered motherhood to be a barrier in their careers, nonetheless, they agreed that motherhood was not easy and did bring significant changes to a woman's life.

Research Requirements and Motherhood as a Choice

One major requirement for sustaining a faculty position in STEM is to continuously publish. All the respondents indicated that this gruelling demand of 'publish or perish' took a huge toll on their mental health. Several of the respondents were anxious regarding their promotions, as they were unable to fulfil their research criteria. The respondents pointed out that publication was just one part of the criteria; they also had foreign collaborations, research grants and conference presentations. Since motherhood required them to be mostly physically present in their homes, travelling for the purpose of research was a major challenge for these respondents. All the respondents stressed the fact that they were wary of applying to prestigious fellowships, which required them to spend months in a foreign country. In particular, even being at the laboratory was difficult, as their children needed their time. Consequently, their research suffered, and they had to endure the success of their colleagues. All the respondents stated that becoming a mother was their own choice, and while that choice gave them immense happiness, it also added stress to their professional lives. They were constantly worried that they would be outperformed by their colleagues.

> I honestly think the rules should be different for mothers. We have to plan so much. Like, for example, if there is a conference at my department, I cannot stay for all the sessions. I have to check on the baby. Whether the baby is eating properly or sleeping properly, it is always the mother's burden. So, I cannot really pay much attention to my research. I feel the promotion norms should be different for mothers. We should be promoted based on our teaching and other roles. Research needs time and thought. It is difficult to devote a substantial amount of time to research with a baby. (Veena, aged 36).
>
> Apart from other things, I have so many PhD students. They need my guidance. In STEM, if you do not have doctoral students under you, then promotion becomes difficult. I also have to ensure that my students get their publications and attend conferences. It's a huge responsibility. The pressure often gets to me, and I feel I am not doing enough as a mother. My child is often with the nanny while I am attending meetings with my students. Hence, my research suffers. One needs time to write a grant and pursue collaborations. I hardly get any time. (Himani, aged 43)

Given that STEM as a field is demanding, it was intriguing for the researcher to understand why these respondents chose to become mothers when they had endured a significant amount of struggle and hardship in pursuing their PhDs, post-docs and then obtaining a faculty position in a prestigious institute. As the researcher posed this question to the respondents, their narratives revealed that they wanted to perform this dual role. Though succeeding in their field was important to them, it was equally important for them to have at least one child. All the respondents, during the interviews, indicated that at that point, having another child was not on their minds. They wanted to do a good job raising one child. Another major reason for having a child was that the respondents wanted to narrate their journey of establishing themselves in the field of STEM to their child. Thus, the respondents emphasized that, in addition to being good researchers in their respective disciplines, they also wanted to raise their child well.

> I am aware that my life would have been much easier if I had remained childfree. In fact, many women are doing that these days to achieve their career goals. But I wanted both. I really wanted to be a mother. I think we women can multitask, and I wanted to prove that I can succeed in spite of being a mother. I wanted to be that change so that other women can also understand that one does not have to sacrifice one role for the other. One can perform both roles with confidence. (Mahima, aged 44)

As these mothers dealt with multiple concerns, their main suggestion was that they should be adequately represented at the level of decision-making. All the respondents highlighted that all the decision-makers were men who lacked any understanding of what women or mothers have to undergo to succeed in the discipline of STEM. In particular, all the respondents unanimously agreed that there is a general lack of respect for mothering and 'women's work' in scientific communities and that women's concerns are often dismissed. All the respondents stated that if their male colleagues started seeing them as equals and started treating them with dignity, then it would be much easier to work for them.

STEM Academia Demands and Institutional Daycares

All the institutes wherein the respondents were working had a daycare system in place. These daycares permitted children in the age group of two to five. The timings of these daycares were mostly from

9:00 a.m. to 5:00 p.m. They were not very expensive, and monthly fees ranged from $36.61 to $61.01. The staff of these daycares were mostly trained personnel (all women), and all these daycares had at least two such personnel. Despite the availability of such daycares, the respondents indicated that they grappled with multiple issues regarding childcare. One of the main grievances expressed by all the respondents was that when their child was sick, they would have to take leave, as they were not allowed to leave their child in the daycare. Talking about the breakdown of the joint family system in India, these mothers indicated that traditionally, in India, women depended upon their mother-in-law or their own mother to raise their children. However, now the situation has changed. Living arrangements in India are undergoing change, and most adult children and their parents or parents-in-law do not cohabit with each other. As a result, several of the respondents had hired a nanny for their children. Though childcare systems were in place, the respondents suggested that they still had to monitor their child from time to time, as neither the nannies nor the daycare personnel paid much attention to their child's well-being.

> The daycare is the only option. At least there, the child has company. At home, the child often gets bored. My main challenge, however, is that the daycare closes at 5 p.m. I am often required to work late at the lab for research. Most of our experiments happen at night, and I am expected to be there at the lab to monitor. Hence, I have hired a nanny who stays with us. Though my husband is there to be with our child, nonetheless, it is always better if a woman is there. She knows the needs of my child. I feel very guilty leaving my child and working late at night, but I do not have a choice. I have to perform as an academic as well as a mother. (Anitha, aged 36)
>
> One of the biggest problems is that I am a part of many committees, and many a time these committee meetings start at 6 p.m. I have to attend them online as I cannot stay any longer at my office. It is also difficult to get anyone to work after 5 p.m. as most of the nannies leave by then. To date, I have not taken a single leave

> for my self ever since my daughter was born. Motherhood is really hard, and my husband is also working, so he cannot really look after the child as he does not understand what to do when my daughter is hungry or sleepy. My parents do not want to live here permanently; they come and go. Hence, I do not have any permanent support. Now, with the pressure to publish, and since our results sometimes take a long time, it is not easy. In fact, I planned my child after I was confirmed in my job and not before that. These days, one has to be very careful as the job market is not very promising. I truly feel that working mothers, in general, have very little support in the country. (Ranjini, age 35)

Despite their tremendous work pressures and motherhood responsibilities, none of the respondents ever thought of quitting their jobs. Being in the labs and able to work in the discipline of their choice gave them a sense of fulfilment. In particular, the respondents stressed that they were PhD holders and faculty members at premier institutes. Their doctoral journeys acted as reminders for them to pursue their STEM careers. Thus, they wanted to make significant contributions to their areas of research as well.

> I am the first PhD in my family. Being a woman, I was expected to marry and have children, but my parents never pressured me. I was a good student and was always told that I should pursue my dreams. For my parents, it is a big deal that I am working at a public organization; hence, I have to strike a balance between both my roles. I am very happy being a mother and holding a permanent position at a prestigious institute. I am also blessed that none of my family members ever ask me to give up on my dreams. My husband is very supportive. So, it is challenging and demanding, but I think that I am very happy doing both my duties.Having said that, I would like to add that I cannot really explore many opportunities. For example, before applying for any fellowship abroad, I have to seriously consider who will take care of our child. Husbands can only do so much, and I have to be here. Thus, it is hard to hold higher administrative positions, and at the moment, I am happy with the current state of things. (Sheila, aged 39)

Apart from their parents, the respondents also mentioned that they had received significant support from their husbands and in-laws. Unlike normative expectations, wherein a woman is expected to give up her job, these women were motivated by their partners to pursue their careers in the field of STEM. Another important factor that the respondents pointed out was that their partners also belonged to the same field. For all the respondents, their partners were either in an equal position or in a higher position compared to their wives. In many cases, the husbands were appointed as heads of their departments, while their wives were still waiting for their turn. Hence, the respondents did highlight that while their husbands were very encouraging, nonetheless, they were in a better position in comparison to their wives. The respondents emphasized that their husbands could travel for research and collaborations, while it was much harder for them to coordinate such activities. None of the respondents blamed motherhood for their not being in positions of authority; instead, the respondents believed that, in the broad discipline of STEM, the leaders will always be men. However, most of the respondents pointed out that, though they are happy balancing their motherhood roles and having a career, they are aware of constraints and that applying for long-term fellowships remains a challenge for them. Consequently, there is a lack of understanding of the issues women, and specifically mothers, undergo. Expecting the state to intervene in appointing women at higher levels, these respondents hoped that their future generations would have it easier.

The main aim of this chapter was to understand the intersections between motherhood and STEM academia in India. In particular, this chapter intended to examine motherhood as a choice and how these women who opt for motherhood navigate their career aspirations in STEM. Given that STEM as a field is particularly dominated by men at the level of decision-making, this chapter chose to explore the position of mothers and their goals and ambitions in STEM research. The narratives from this chapter revealed that, though the respondents had taken a considerable amount of time to finish their PhDs and to obtain a faculty position in a prestigious institute, nonetheless, it was their own individual decision to become mothers as they wanted to perform both roles, that of a STEM academic and a mother. Additionally, these respondents highlighted that they had not been pressurized by their husbands or their parents/parents-in-law to embrace motherhood. In fact, it was their own choice to have a child. All the respondents were aware that having a child would lead to career breaks and many other restrictions in terms of their research; nonetheless, motherhood gave them immense joy. Recent studies have noted that the idea of motherhood is undergoing change in urban India, with women opting out of motherhood by remaining childfree for career aspirations (Bhambani and Inbanathan, 2018). The findings of these studies suggested that the need to focus on one's self-development, personal aspirations and desires are the key reasons for women to not opt for motherhood. Going beyond their reproductive roles, women in India are now choosing to nurture their own ambitions. In contrast to these studies, the narratives of this chapter are striking, wherein women, in spite of investing their energy and time in attaining a PhD in STEM, chose to become mothers and seek contentment by fulfilling dual roles.

The main concern expressed by these respondents was their lack of representation in top leadership. As indicated by the respondents, the rules for promotion should be separate for mothers, as women would want to pursue both roles. Owing to similar research criteria for all faculty members in STEM, the respondents stated that they were at a disadvantage. As the narratives suggest, many of the respondents cannot travel or spend long hours at their labs because of their motherhood duties. Additionally, administrative responsibilities take up a lot of their time, and they are unable to allocate adequate time and effort to their research. Kaur and Kapoor (2022), in their recent paper, found that women in STEM are often under-represented in leadership positions, as women are still viewed as caregivers and nurturers. Through their analysis, Kaur and Kapoor

(2022) suggest that women are often offered administrative roles in human resources and student welfare responsibilities, wherein they can utilize their caregiving and nurturing skills, which are considered to be natural attributes of womanhood. Echoing the findings of this paper, the respondents of this chapter also indicated that while they are part of many committees, they are hardly represented at the top level. Additionally, as the respondents pointed out, since the higher-level administration mostly consists of older men, they always compared women faculty, specifically women faculty who are mothers, to their wives who have sacrificed their careers for their husbands and children. As these women are contesting the normative narrative of being a mother, they expect that research norms and publication requirements should be more relaxed for them, as they cannot engage in many collaborations or research grants owing to their motherhood commitments.

The final finding of this chapter delineated that the husbands of the chapter respondents were quite supportive and motivated their wives to pursue their careers. Though all the respondents relied on institutional care for their children, nonetheless, they were never pressured to quit their jobs. Not only the husbands, but there was also no pressure to quit from the parents or the parents-in-law either. Despite this encouragement, the respondents did highlight that their husbands were in a much better position as compared to them, as their husbands had held or were holding leadership positions, while the respondents were unsure if they would ever be offered such positions. As outlined by Doerr (2022: 710–23), women are increasingly fearing motherhood as they believe that post motherhood their own ambitions and aspirations will not achieved. Thus, to support mothers in STEM academia and for other women in STEM academia who might nurture the idea of embracing motherhood, it is important to devise policies that will enable women to maintain a balance in both their roles.

As discussed by scholars of feminist science studies, it is important to delineate the complexities of science as a discipline, as well as explore the many intersections of social constructs such as race, gender and caste. Adding to this discussion, this chapter highlights how the experience of STEM mothers is unique and needs more in-depth analysis as a separate category to reduce the inequalities faced by mothers in STEM academia and laboratories.

One of the major concerns shared by all the respondents was that STEM academia should have separate rules and policies for mothers. As indicated by the respondents, the lack of female representation at the level of leadership in the majority of STEM institutes is the key reason why the requirements of mothers are never truly understood. As most committees consist of older men, their lack of empathy towards working mothers is a major challenge encountered by all the respondents. To address this concern, it is important that an adequate number of women are appointed to decision-making bodies. The other significant issue raised by the respondents was that having similar research criteria coupled with equal administrative workloads for mothers is unfair, as mothers have to balance their work and childcare responsibilities. This issue should be noted, and there should be separate promotion policies in place that are applicable only to mothers. In particular, women should not be expected to publish during their maternity leave, as that leave is given for women's bodies to recuperate after giving birth, and it is not a paid leave period. Narratives from the chapter revealed that women chose to play dual roles, that of an academic and a mother. As emphasized by the respondents, motherhood was their individual choice, and it was not enforced upon them. For women to be able to pursue their careers as well as perform their motherhood duties, it is important that STEM research and academic organizations have flexible timings and better leave policies. In most cases, as this chapter found, mothers have to sacrifice their own health and use their leaves for childcare. In particular, as parenting is mostly seen as a mother's responsibility, most mothers apply for leaves judiciously. They prioritize their child's health over their own physical and mental health. However, with additional leaves, mothers will not have to choose between their own health and their child's health.

The narratives of this chapter suggest that the family as a unit is changing in urban India, with women receiving support from their family members and husbands to retain their faculty positions in a premier STEM institute. In addition to the family, the Government of India has also noted the gender inequality in STEM academia and has floated various schemes that will encourage more women to pursue a career in STEM. Schemes such as the Women Scientists Programme and SERB-POWER (Promoting Opportunities for Women in Exploratory Research) have been launched to motivate more women to enter and engage in STEM research. Adding to these schemes, if some amount of quota is implemented by the state to appoint women to positions of authority, it might enhance the well-being of women and mothers in STEM academia.

Despite inherent gender disparities, motherhood in STEM academia remains under-researched in India. This chapter hopes to contribute to the gender studies scholarship of India and encourage scholars in this field to further explore questions that will shed light on the intersections between caste, religion and gender in STEM organizations in India.

References

Beutel, Ann M. and Cyrus Schleifer. 2021. 'Family Structure, Gender, and Wages in STEM Work'. *Sociological Perspectives* 65(4): 790–819. https://doi.org/10.1177/07311214211060032.

Bhambhani, Chandni and Anand Inbanathan. 2018. 'Not a Mother, yet a Woman: Exploring Experiences of Women Opting out of Motherhood in India'. *Asian Journal of Women's Studies* 24(2): 159–82.

Budig, Michelle J., Joya Misra and Irene Boeckmann. 2012. 'The Motherhood Penalty in Cross-National Perspective: The Importance of Work–Family Policies and Cultural Attitudes'. *Social Politics* 19(2): 163–93. https://doi.org/10.1093/sp/jxs006.

Chadha, Gita and Asha Achuthan. 2017. 'Feminist Science Studies: Intersectional Narratives of Persons in Gender-Marginal Locations in Science'. *Economic and Political Weekly* 52(17): 33–6.

Correll, Shelley J., Stephen Benard and In Paik. 2007. 'Getting a Job: Is There a Motherhood Penalty?' *American Journal of Sociology* 112(5): 1297–338.

Doerr, Katherine. 2022. 'Flying under the Radar: Postfeminism and Teaching in Academic Science'. *Gender, Work & Organization* 31(3): 710–26. https://doi.org/10.1111/gwao.12922.

Dunn, Marianne, Margo Gregor, Simone Robinson, Anthony Ferrer, Devynn Campbell-Halfaker and Javier Martin-Fernandez. 2022. 'Academia during the Time of COVID-19: Examining the Voices of Untenured Female Professors in STEM'. *Journal of Career Assessment* 30(3): 573–89. https://doi.org/10.1177/10690727211057441.

Fuller, Sylvia. 2018. 'Segregation across Workplaces and the Motherhood Wage Gap: Why Do Mothers Work in Low-Wage Establishments?' *Social Forces* 96(4): 1443–76. https://doi.org/10.1093/sf/sox087.

Fulweiler, Robinson W., Sarah W. Davies, Jennifer F. Biddle, Amy J. Burgin, Emily H. G. Cooperdock, Torrance C. Hanley and Maren Ziegler. 2021. 'Rebuild the Academy: Supporting Academic Mothers during COVID-19 and Beyond'. *PLoS Biology* 19(3): e3001100. https://doi.org/10.1371/journal.pbio.3001100.

Haraway, Donna. 1987. 'A Manifesto for Cyborgs: Science, Technology, and Socialist Feminism in the 1980s'. *Australian Feminist Studies* 2(4): 1–42. DOI: 10.1080/08164649.1987.9961538.

Haraway, Donna. 1992. 'A Cyborg Manifesto: Science, Technology, and Socialist-feminism in the Late Twentieth Century'. In *Simians, Cy- borgs, and Women: The Reinvention of Nature*, 149–82. New York: Routledge.

Hsu, Chen-Hao. 2021. 'Parity-Specific Motherhood Penalties: Long-Term Impacts of Childbirth on Women's Earnings in Japan'. *Advances in Life Course Research* 50: 100435. https://doi.org/10.1016/j.alcr.2021.100435.

Kahn, Joan R., Javier García-Manglano and Suzanne M. Bianchi. 2014. 'The Motherhood Penalty at Midlife: Long-Term Effects of Children on Women's Careers'. *Journal of Marriage and Family* 76(1): 56–72. https://doi.org/10.1111/jomf.12086.

Kaur, Ravinder and Tanya Kapoor. 2022. 'Securing the Leak, Shattering the Glass Ceiling'. *Seminar.*

Keller, Evelyn Fox. 1982. 'Feminism and Science'. *Signs: Journal of Women in Culture and Society* 7(3): 589–602.

Keller, Evelyn Fox. 1987. 'The Gender/Science System: Or, Is Sex to Gender as Nature Is to Science?' *Hypatia* 2(3): 37–49.

Masoodi, Ashwaq. 2016. 'Where Are India's Female Scientists?' *Mint*, 23 October. Available at: https://www.livemint.com/Politics/N0j16kL2ZAzk94ssOKoTdK/Where-are-Indias-female-scientists.html.

Morgan, Laurie A. 1998. 'Glass-Ceiling Effect or Cohort Effect? A Longitudinal Chapter of the Gender Earnings Gap for Engineers, 1982 to 1989'. *American Sociological Review* 63(4): 479–93. https://doi.org/10.2307/2657263.

Purcell, David, Kelly Rhea MacArthur and Sarah Samblanet. 2010. 'Gender and the Glass Ceiling at Work'. *Sociology Compass* 4(9): 705–17. https://doi.org/10.1111/j.1751-9020.2010.00304.x.

Ray, Debapriya and Tannistha Samanta. 2020. 'Motherhood Penalty? Examining Gender, Work, and Family among Science Professionals in India'. *Navigating Careers in the Academy: Gender, Race, and Class* 3(1): 37–51.

Sahoo, Priyanka. 2021. 'Few Women on Top: Time to STEM the Gender Gap'. *Hindustan Times*, 15 June. Available at: https://www.hindustantimes.com/cities/others/few-women-on-top-time-to-stem-the-gender-gap-101626634461832.html.

Simon, Marsha and Jean Swindle. 2024. 'Daring to Be a Mother: A Case Study on Being Black, Being Pregnant, and Being a STEM Doctoral Student'. *Educational Studies* 60(5): 551–73.

Subramaniam, Banu. 2009. 'Moored Metamorphoses: A Retrospective Essay on Feminist Science Studies'. *Signs: Journal of Women in Culture and Society* 34(4): 951–80.

Thébaud, Sarah and Catherine J. Taylor. 2021. 'The Specter of Motherhood: Culture and the Production of Gendered Career Aspirations in Science and Engineering'. *Gender & Society* 35(3): 395–421. https://doi.org/10.1177/08912432211006037.

The Wire. 2018. 'Tenure Clock Pause, Daycare Revolution: Weapons to Fight Sexism in Science'. Available at: https://thewire.in/women/tenure-clock-pause-daycare-revolution-weapons-to-fight-sexism-in-science.

Yu, Jia and Yu Xie. 2018. 'Motherhood Penalties and Living Arrangements in China'. *Journal of Marriage and Family* 80(5): 1067–86. https://doi.org/10.1111/jomf.12496.

8

Navigating Stigma: Case Studies of Childfree Women in Indian Cities

Rhea Jaikumar Menon

Introduction

Fertility and birth rates have steadily declined in India over time. In the 1960s, India's total fertility rate (TFR) was 5.9 children per woman (United Nations Population Division 2024). Over fifty years later, India's TFR has fallen to 2.0 children per woman, dropping below the replacement level of 2.1 for the first time (IIPS and ICF 2021: 111). According to the fifth round of the National Family Health Survey [NFHS-5], conducted in 2019–21, urban India reports an even lower TFR of 1.6 compared to 2.1 in rural areas (IIPS and ICF 2021: 123). The persistent declining trend of fertility rates in India indicates an attitudinal shift in childbearing preferences. Household-level data reveals that families have nuclearized and couples choose to have fewer children, limiting themselves to one or two. However, national statistics have overlooked a key emerging demographic trend – the growing number of childfree people in India.

Childlessness spans a spectrum of choices and circumstances. Childlessness may be voluntary or involuntary or occupy a vague liminal position in between, thus forming a 'continuum of childlessness' (Berrington 2017: 58). An individual might be involuntarily childless due to biological infertility or social circumstances, such as not finding a partner yet. On the other end of the spectrum, a person may intentionally forgo parenthood to be voluntarily childless for various reasons, regardless of their biological fertility or relationship status. Over time, the word 'childfree' emerged as a preferred term for voluntary childlessness because of its positive and affirmative connotations (Basten 2009: 9). In India, childfree people represent a small yet growing section of society whose perspectives reflect the nature and trajectory of socio-demographic change in the country.

National statistics do not adequately measure the rising number of childfree persons in India. The NFHS only documents the fertility and childbearing preferences of married couples, and thus, single childfree people are statistically neglected. According to NFHS-5, 5 per cent of married women and 36 per cent of married men in India are childless and do not express a desire for children. No data about the rationale for their choice is recorded. Thus, the NFHS and other national surveys fall short of capturing the size of the emerging demographic phenomenon. Instead, the growth of the childfree segment in India is highlighted through news, pop culture and the emergence of online childfree communities.

The childfree movement has gained increasing visibility in India, particularly in mainstream media. News outlets and online publications have generated awareness by sharing personal narratives from childfree men and women in Indian society. Social media sites like Facebook and Reddit host platforms for forums and groups dedicated to the issues, perspectives, and experiences of childfree Indians. The childfree trend is a pan-India phenomenon, but it is most evident in urban areas where the discourse and digital activity are focused. Lastly, the childfree choices of celebrities and representation in popular culture have strongly contributed towards the general awareness of the childfree phenomenon in India.

The term 'childfree' is not always explicitly used. Notable Bollywood celebrities like the actor Akshaye Khanna (*Hindustan Times*, 2019) and the actress Vidya Balan have stated that they do not want children and are widely assumed to be childfree. Female personalities are more frequently probed on their intention to have children than their male counterparts. Vidya Balan, married for over eleven years without children, becomes the target of pregnancy rumours whenever her weight fluctuates. Unlike her male co-stars, she is often asked about pregnancy and children during press events (*Filmfare*, 2019). This differential treatment is not only limited to public figures. Within a childfree couple, the wife may be subjected to more questioning, judgement and unwarranted advice from others compared to her husband (*Times of India*, 2017). Despite their growing presence and numbers, childfree people in India, especially women, face stigmatisation for deviating from social norms that prescribe childbearing.

The gender disparity and the fixation with procreation reveal the pronatalist patriarchal undertones in a collectivist, tradition-bound society like India. However, these cultural tendencies have not hindered the rise in childfree numbers, especially in urban India. Spurred by socio-economic development and attitudinal shifts in childbearing preferences, the rapidly growing minority calls for attention. With higher educational attainment, more work opportunities and increasing access to contraception and family planning over time, Indian women can choose to have fewer children, if at all. As birth rates continue to decline, India will face a population ageing crisis in the coming years as the social burden of elderly needs expands. If the childfree trend continues to be overlooked, stigmatization will also persist, further worsening the social condition of childfree people in India, especially women.

The Stigmatization of Childfree Women in India

Erving Goffman's definition and theory of stigma can be applied to the decision to be childfree. The term 'stigma' refers to an 'attribute that is deeply discrediting', such that it alienates the attribute – holder from other persons in the category they might have belonged to, or it reduces a 'whole and usual person to a tainted, discounted one' (1963: 3). The term needs to be understood relationally because what is discrediting to one individual might be acceptable to another.

Unlike an evident characteristic like gender or race, being childfree is an invisible yet discrediting attribute within heteronormative or pronatalist environments. It is only upon revelation that childfree people may face devaluation and discrimination (Goffman 1963: 73). Thus, childfree individuals may deploy stigma management strategies to mitigate adverse responses from others. One strategy is 'passing', where younger childfree individuals state that they will have children in the future. Another strategy is 'identity substitution', where a childfree person will falsely claim to be infertile. A third strategy relates to 'primary control coping', where a childfree person may turn stigmatic assumptions of selfishness back towards parents and their rationale for having children (Park 2002: 39; Verniers 2020: 113). In India, where society is governed by heteronormative and patriarchal notions, researchers have launched novel forays into understanding the stigmatization of childfree people.

Under patriarchal conditions, the institution of motherhood is culturally naturalized within Indian women, thus stifling their potential and agency. Amrita Nandy coined the term 'mater-normativity' to describe this socio-cultural phenomenon (2017: 63). Nandy (2013; 2014) highlighted how the thoughts and actions of those who subvert mater-normativity, including childfree women, are questioned intensely and often stigmatized. They are deemed 'selfish' or 'abnormal' for rejecting motherhood. Other derogatory labels include "lazy" and "irresponsible" (Nandy 2013: 56). However, Nandy's interviews with childfree women contradicted these descriptions. The women interviewed valued self-determination through social work and professional pursuits (Nandy 2013: 56). Though they opted out of motherhood, most were still engaged in other forms of caregiving, like caring for younger siblings or elderly parents (Nandy 2013: 56). Instead of leading lonely lives, they cultivated their social circles, even reimagining family to include friends (Nandy 2013: 56). Some participants loved children and expressed fleeting curiosity or desire for motherhood, yet chose to remain childfree after deliberation. In online spaces, they critiqued biological essentialism, maternal norms, taboos and performativities (Nandy 2014: 3). However, they did not condemn motherhood nor criticize mothers. The narratives of the women interviewed simply revealed new and multiple ways of navigating identity and being in India outside of culturally mandated motherhood.

In a similar undertaking to Nandy's research, Chandni Bhambhani and Anand Inbanathan (2018) also conducted fieldwork interviews and analysed the reasons childfree women forgo motherhood. There were many significant similarities and intersections with Nandy's research in terms of the rationale behind the childfree decision and the stigmatic experiences of the women. In the Bhambhani and Inbanathan study, childfree women additionally challenged the childbearing rationale of existing and aspiring parents, suggesting that the choice was for 'selfish' reasons like fulfilling the parents' emotional needs or having a caretaker in old age (2018: 168). The women also stated that the burden of childcare fell heavier on women under patriarchal conditions, and that was an additional deterrent towards motherhood (2018: 162). Bhambhani and Inbanathan reached a similar conclusion as Nandy – through their exercise of autonomy and agency, childfree women can reimagine and redefine womanhood beyond motherhood in India.

As womb-bearers in a traditionally patriarchal society, childfree Indian women bear the greater burden of stigmatization. However, in the face of stigma, they respond through debate and defiance. They argue that their stance is neither selfish nor irresponsible but rather the opposite. Citing concerns such as lifestyle choices, financial stability, environmental issues and genetic risks, they assume responsibility by opting out of parenthood. Instead of succumbing to social pressures and following norms mindlessly, childfree individuals deliberate carefully over the decision to have and raise children before arriving at their conclusion (Firstpost, 2022). In India, childfree women defy the cultural mandate of motherhood and construct their own life paths independent of social norms and expectations.

Despite its growing significance, academic research on the childfree phenomenon in the Indian context remains limited and nascent. To mitigate demographic crises and social disharmony, the narratives of childfree Indians must be acknowledged and further understood, especially those of childfree women. Research has primarily focused on the reasons behind forgoing motherhood. The post-decision impact of the childfree choice on personal, professional and social domains has also not been studied in detail. The role of the childfree identity and consequent stigmatization in shaping the lives of women in India remains a largely uncharted territory, which this chapter intends to explore. This chapter reveals the nuanced narratives of stigmatization among middle-class childfree women in urban India. In urban areas, childfree women are empowered by neoliberal values and socio-economic freedoms, but they still experience stigmatization from loved ones and wider society. Through five case studies, this chapter explores why certain women in urban India forgo parenthood, the impact of the childfree decision on their interpersonal relationships, and how they manage the stigma that emerges from defying traditional social norms.

Prospective participants for the study were approached on the social media applications Reddit and Discord via forums dedicated to childfree people in India – r/ChildfreeIndia on Reddit and the Childfree India Discord server. The sample was limited to childfree women between 25 and 45 years old residing in Indian cities. Five participants were chosen to share their narratives in exhaustive detail and shed light on the lives of childfree people in India. The insights derived are not meant to generalize the entire childfree Indian population but instead demonstrate the diversity of narratives and scope for further research and attention.

The interviews were based on semi-structured questionnaires, which resulted in five comprehensive case studies. Important and recurrent keywords were noted and recorded to be organized into narrative themes later in the analysis. Thus, a grounded understanding of childfree women in India was developed without relying upon pre-existing theory and research. During the analysis stage, the stigma management strategies employed by the women were compared against Goffman's theories of stigma to understand if they hold true among childfree people in the Indian context.

Here are the basic demographic profiles of the five women interviewed for the study.

1. Anushka is a 26-year-old bridal hairstylist based in Jaipur. She is single and not religious. She holds a bachelor's degree in psychology.
2. Divya is a 26-year-old childfree woman living in Delhi. She is single and not religious as well. She holds a B.Tech degree but is currently not working and is preparing for the civil services entrance examination instead.
3. Aisha is a 28-year-old content developer working in Bangalore. She is Muslim and has been married for over two years. She has an M.Tech degree in Chemical Engineering.
4. Nidhi is a 30-year-old doctor in Bhopal. She is unmarried but has been in a long-term relationship for over seven years. She is agnostic and holds two master's degrees in different medical specializations.
5. Meera is a 40-year-old childfree woman who recently relocated to Dubai from Bangalore. She owns a consultancy, where she serves as COO. She has been married for over thirteen years. She is an atheist and holds a master's degree.

Reflections from Childfree Women in Indian Cities

Reasons for Being Childfree

Each interviewee had at least two reasons for being childfree – a primary reason, which is the main reason behind their decision, and secondary reasons, which are contributing factors. The primary reason for each respondent varied, but multiple respondents shared secondary reasons in common.

Anushka is the youngest interviewee and shared several reasons behind her decision to go childfree. She cited the fear of pregnancy as her primary reason and said, 'Pregnancy has always been a scary thing for me.' She is scared of pain, hormonal imbalances, postpartum depression and 'a thing living inside you'. Though she did not label her fear as a phobia, Anushka might actually have primary tokophobia, a morbid fear of childbirth without any previous experience of pregnancy. She is also not keen on pursuing motherhood through means like adoption or surrogacy either. She believes that 'the world is not a great place right now' with struggles like economic depression, war and hunger. According to Anushka, 'nobody in this world [including herself] is happy', and her potential child would also not be happy, regardless of how she raises them. She does not want to give birth to a child and bring them into such a world without their consent.

Anushka stated that 'kids are really expensive', referring to the costs of education, medical care, food and others. Instead, she would rather 'do something better' with the money, like 'travel the world'. Anushka also brings up the concern that 'after spending so much money on a kid, [she] can't guarantee the kid is going to be good'. She phrased her concern accordingly:

> What if he turns into a terrorist? What if the kid turns into a very shitty child? Or just an asshole in general? Or does something really bad to somebody? I don't want to raise a person like that and then regret later, thinking I shouldn't have done it.

Anushka's final reason for being childfree is her career as a bridal hairstylist. Her occupation requires her to travel out of town frequently, and she does not want to 'burden' her prospective partner or family members with childcare. She has put 'years of effort' into her career and does not want to 'compromise' on it by having a child.

Divya's primary reason for being childfree is her lack of 'motherly instinct,' coupled with no desire for motherhood.

> I don't have a motherly instinct. I don't feel any sort of emotion when I look at a child. I don't want a child to be mine. Back in college, all of my friends liked kids, but to me, kids were the biggest burden I could think of. When I looked at the women around me, kids seemed severely limiting to everything they did. As I grew older, I realised I don't have any maternal feelings. It's not like I am a mother right now, but I can't even imagine myself getting to that point.

Regarding the lifetime responsibility of caring for a child, Divya confided that she does not 'want that kind of situation' for herself. Divya also shares a similar opinion on the world with Anushka. She said, 'The rest of the world is burning; why should I bring another child [into this].'

Aisha is also not keen on caring for a child in addition to herself, and this is her primary reason for being childfree. Similar to Anushka, she wants to prioritize her personal pursuits over motherhood. She was the only religious participant in the study, but her religion is not a contributing factor towards her decision to be childfree.

> I think life is, anyway, too much to handle, and handling somebody else will be another [struggle]. I have some of my own plans that I want to pursue. And if I do have kids, I'm pretty sure I won't be able to [raise them]. I might not hate [the child], but I will regret the decision later, and I know that for a fact.

Like the others, Nidhi did not want the responsibility of caring for a child and expressed her lack of desire for motherhood as her primary reason for being childfree. She also demonstrated a rational approach towards decision-making.

> I look at all decisions that we have to make as being based on our interests, and you really need to want [to become a mother] in order to do anything like that. It did not make any sense for me to make a decision just to please somebody else. [Motherhood] is also like your career or a hobby, you know—it is for a lifetime. It's not like you just have a kid and then decide, 'I'm not liking it, where do I return it?' It's not returnable. I knew this early on. I think I've always been childfree somewhere in the back of my mind. I never had that idea that maybe I'll have kids someday.

Meera's reasons for being childfree differ from those of Divya and Nidhi, but her thought process is rational and thorough, like theirs.

> I can't think of a reason to have a child, very honestly. I think when I was a kid, around seven or eight years old, I wanted to adopt a child, and I've never wanted to have my own children. I don't come from a stable background. I come from a lot of abuse, a lot of nonsense that children should not go through. I was about twenty-one or twenty-two when I walked away; I walked out of home.

Meera did not specify which reason is her primary one, but it is evident that her childhood abuse serves as a strong deterrent to the pursuit of motherhood. Being childfree is a rational conclusion the participants arrive at after extensive consideration of their personal circumstances and external factors. Though their reasons for being childfree varied slightly, the women shared a lot in common. They decided to be childfree in light of their best interests, but they are not 'selfish,' as often stigmatised. Their choice does not inflict harm and is not made at the expense of others. They express concern over the state of the world and the well-being of their hypothetical children. In scholarship, concern for the environment has been cited as a reason many people decide to be childfree. However, for these five participants, environmental concern and climate change were not among their top reasons. This highlights the scope for further research into attitudes towards climate change among childfree people in India compared to other countries.

Impact of the Childfree Decision on Interpersonal Relationships

The decision to be childfree can only impact potential and current relationships if it is disclosed. Thus, the participants could choose to disclose or selectively disclose their childfree status to manage its impact and possible stigmatization. They are mostly forthcoming about their decision to be childfree, but to different degrees of openness depending on the type of interpersonal relationship. The women are:

1. Open about their childfree status to prospective or current romantic partners,
2. Moderately open to colleagues, friends, and wider family members, and
3. Selectively open to parents and close family members.

Having children is a normative event within romantic relationships, and being childfree must be discussed to ensure compatibility and alignment of values and life goals. Regarding the second point, the women only shared their stance when questioned about their lack of children but did not actively try to conceal it. In contrast, they were more open yet selective about disclosing their childfree decision to family relatives.

Parents and close family members are strongly concerned with the life choices of individuals in a collectivist, traditional society like India. Participants have to navigate the issue of selective disclosure carefully. For instance, they may be open with their parents but hide their position from their mother-in-law. Mother-in-law and daughter-in-law relationships can be contentious for many reasons, and the latter's childfree status may worsen or strain the relationship further. The impact of the childfree decision on interpersonal relationships of childfree women varies significantly on a case-by-case basis, but women maintain agency over the disclosure of their choices.

For decidedly childfree women, not being childfree is a potential dealbreaker, and the dating pool becomes significantly smaller and less accessible. Without approaching people via a childfree community or mutual connections, it is difficult to determine if someone is childfree before dating them. Anushka shared that being childfree has had a noticeable impact on her dating life. Dating as a childfree woman has not been the smoothest experience for Anushka.

> I went on a date with this one guy who was adamant that he wanted a relationship and not something casual. So, we discussed it, and I told him that I'm childfree. This guy was great, amazing. I'm still friends with him, but it did not work out simply because he was not childfree, and I was. The moment I got to know that he wasn't childfree, I took a step back and could no longer consider dating him.

Not being childfree is not necessarily a dealbreaker when the partner is ambivalent or undecided about having children. Divya has had two boyfriends, and neither was childfree when she started dating them.

> My last boyfriend was agnostic about it. He didn't care if I had a kid or not. Sometimes partners are agnostic about it, and they're okay with being childfree, with being with you. But it will be hard for them to explain this concept to others, and I kind of don't blame them. Both of us knew we were going to run into problems convincing his mother. He was willing to say, 'I'm completely infertile, she's not, and we're not willing to do any procedure or anything.' This was how he wanted to tackle his mother. I, on the other hand, said, 'I don't mind taking the blame for it. Why do you have to say you're infertile when I can happily tell them this is our decision because this is my choice?'

Being childfree did not contribute to Divya's most recent break-up. However, her conversation with her boyfriend revealed a stigma management strategy that can be used by childfree individuals – identity substitution. He believed that pretending to be infertile is better than disclosing being childfree because the former would be easier to accept by his mother. He was also willing to assume the burden to make the lie more acceptable.

Early disclosure prevents a doomed relationship from progressing when there is a difference in ideals and values. Relationships can only be healthily sustained when both partners agree on being childfree. Aisha and her husband have some disagreements over her childfree status. Aisha has been with her husband for over five years and married for the past two years. At the beginning of their relationship, she told him that she had two medical conditions, dysmenorrhoea and hypothyroidism, which would make it difficult, if not impossible, to endure pregnancy. He asked her about adoption, but Aisha made it clear that she would remain childfree, and he accepted it. Aisha suspects that he has not entirely accepted it, though.

> Lately, I think that when he sees his peers having kids, he wants to have a kid as well. I told him that he can adopt, but I won't be a part of the kid's life. If you [her husband] are ready to do 100% of it, then go ahead. But you can't make me do it, simply because I can't, and I don't want to. Now, if you are not ready to give your 100%, then you are not fit to have a kid, because there are people who are single parents, and they are doing it. You cannot adopt a child and put the responsibility on me.

Aisha admits that their differing positions, in addition to questions and mounting pressure from other family members, have caused some strain on their marriage. However, Aisha believes that once he deliberates more on 'the rising costs of living' and other factors, he will agree to being childfree again.

Meera presented a unique take on the impact of being childfree on her marriage. Married for thirteen years and dating for five years before, Meera said that being childfree actually benefitted her marriage.

> I would say being childfree is one of the cornerstones of our relationship, in addition to loving animals more than we love ourselves. I think we are able to be around each other more and

> love each other more. We genuinely want to spend more time with each other simply because we have chosen to be childfree. People who come from a very traumatic childhood tend to be more often childfree than people who don't, and both my husband and I had very traumatic childhoods. That's what attracted us to each other in the first place, and we've known each other for a really long time before we dated. Neither of us had the opportunity to be a child and to explore ourselves and our world when we were children. We were forced to grow up before our time. We raised ourselves; we don't want to raise another person.

For single women like Anushka and Divya, the childfree choice restricts the dating pool and makes early disclosure almost mandatory to avoid wasting time and effort on prospects that will not materialize. For childfree women in committed relationships and marriages, being childfree will not cause any problems if there is complete acceptance and consensus between the partners.

Unlike romantic relationships, childbearing preferences are not a constitutive element of interpersonal relationships with colleagues. Ideally, personal life choices should not have any bearing on the work environment. However, disclosure of a non-normative choice, like being childfree, can lead to stigmatization in the workplace. All the participants stated that their childfree status had not affected their relationships with colleagues. The women do not voluntarily bring up their childfree status; however, if the topic is raised, they then reveal their position. They are frequently met with questions and debate. These responses include: 'Why are you childfree?,' 'You will change your mind in the future,' 'Who will take care of you when you are old?' and other common statements that most childfree Indians hear regularly. There is some stigmatization in terms of being dismissed or not taken seriously, but none of the participants claimed that this had a significant impact on their career and professional relationships.

Owing to her unique work environment, Anushka finds herself having to disclose that she is childfree more frequently than most.

Conversation topics around marriage and childbearing are commonly raised in the wedding environment, and Anushka shares her decision to be childfree. Despite her clientele being brides and possibly future mothers, Anushka does not face any backlash for her words. She and her clients and colleagues are only together in a professional capacity, and neither party crosses personal boundaries. Anushka only recalls one incident that stood out to her:

> There is this one colleague who is also sort of a good friend. She has a kid, and we were just sitting in the car talking about it. I told her that I don't want to have kids in general, so she gave me a very weird reaction at that point – a visible change in facial expression. I thought, 'Why are you [the colleague] giving me a weird reaction?' I didn't say it out loud, but I could see it right in front of me. I ignored it because she had a kid herself, so maybe she found it offensive. I don't really care about it, honestly, because we meet each other only when we are at work.

The incident with Anushka's colleague serves as an example of how one's personal choice to be childfree can be perceived as an offensive attack or a prejudiced judgement towards others who are not. The misinterpretation of words and the difference in values can inadvertently create hostility in relationships.

Almost all the women said being childfree gave them more flexibility with geographical mobility and career choice. As a bridal hairstylist, Anushka travels frequently across India to serve her clientele. She enjoys her career and lifestyle and sees having children as an impediment. Aisha said being childfree means being free to choose 'where I want to work and what work I want to do'. Nidhi has always been ambitious, academically and professionally. She is a doctor with master's degrees in two specializations and has been able to shift her medical career from one speciality to another. She remarks that it would not have been possible if she had a child, and as someone who enjoys her career, she still does not want a child. Meera has a lengthy resume and has enjoyed much professional success. She has 'often been the first female employee of most of the organizations' she has worked at. She proudly says that she has put in the same hours and the same amount of effort as her male colleagues. In return, colleagues often appreciate the dedication of the childfree participants to their careers.

Similar to workplace relationships, childbearing is not a feature of platonic friendships. However, in some cases, differences in values can adversely impact friendships. Among the interviewed women, the ones below thirty years old – Anushka, Divya and Aisha – shared that they have not told most of their friends that they are childfree or do not want children in the future. This is because it 'does not come up' and 'is not a point of conversation'. The friends who are aware of their choice are unbothered or understanding of it. Nidhi and Meera, who are thirty and forty years old, respectively, shared different accounts. Nidhi highlighted how being childfree is straining her friendships.

> It is starting to become an issue because a lot of my friends have kids. I want to hang out with them, but when I call them, they're pretty much busy with their kids, projects, and stuff like that. I look at them, and I think to myself, wow, I don't have to do any of that. It feels good, but it's also lonely because there's no one to hang out with.

Meera is a strong advocate for being childfree and prefers the company of pets over children. Her views have influenced her friendships and shaped her social circle.

> The funny thing is, I also attract a lot of women who have chosen to be childfree. A lot of my friends are childfree. Some of them are married, but surprisingly, that's a very small percentage of people. It's always nice to know, when you're close to a person, where they stand on this issue [of being childfree]. I'm just not comfortable around children. My relationship with people, once they've had kids, dwindles, and it becomes non-existent. I have no desire to be somebody's godmother, crazy aunt, cool aunt, or whatever the other stereotypes are.

At a younger age, people are in the same life stages and have similar priorities, with having children further down the list. However, values diverge as people approach their thirties or get married. Some people intend to become parents, while others wish to remain childfree. Individuals are drawn to those they have more in common with. Thus, the compatibility between childfree and non-childfree friends declines in some instances, especially when the former does not want to be around children much. Apart from Meera, all the other women said that they do not know other people who are childfree in their daily lives, since it is a minority stance that often remains undisclosed. Over time, this leads them to feel lonely and not understood, and this is why they are drawn to online childfree communities like those on Reddit and Discord.

Regarding family, all five interviewees are open about being childfree with their parents and have received mixed responses. Anushka's and Divya's parents undermine their childfree positions with dismissive statements like 'everyone has to do it [have children],' 'you will change your mind in the future,' 'you will need someone when you are old,' and other counterarguments. Coincidentally, Anushka and Divya have cited their mothers as their push towards being childfree, unbeknown to them. Anushka recalls an argument with her mother when the latter tried to eat a roti that was misshapen and cold since it was the first one of the batch.

> I literally yelled at her, telling her that this is not okay – you [her mother] cannot keep compromising on your health and on your food just because you want your kids to eat good food. You cannot keep doing that to yourself.
>
> [To interviewer] I can't compromise so much. I won't leave my food and feed my kids; instead, I'll eat more. I'm that kind of person. If I get two ice creams, I would rather eat both of those ice creams than let my kid eat them. I cannot share my food. I don't do it with my brother. I don't do it with my mom. I don't do it with anybody. I cannot do it with my kid either. I'm just saying the kind

> of compromises these moms are supposed to make—I can't do those compromises. That's too much for me.

Anushka does not want to curb her freedom and well-being, and Divya echoes a similar sentiment.

> If I have a child, I will have a lot of duties, and I will have to leave my career. I cannot leave my career; it's just not something I will sit at home for. I will not be like my mother. I can't be the same person.

Anushka and Divya strongly reject the idea of having children because they do not want to compromise on their goals and ambitions. Being a working mother is not appealing to either of them, as well. Instead, they implicitly reject the experiences of patriarchal motherhood they have observed with their mothers.

The partnered participants of the study – Aisha, Nidhi and Meera – are open about their decision to be childfree, and their parents accept it as well. Being married or in long-term relationships appears to soften the backlash since it is less deviant relative to being both single and childfree. Aisha's parents are aware of her medical conditions and do not want her to bear children at the risk of her health. Nidhi's parents were worried that she would be lonely as she gets older, but their concern has been abated because of her long-term romantic relationship. Meera has a complicated relationship with her parents, but her parents begrudgingly accept her childfree status because Meera would 'easily walk away from them [the parents] otherwise'. However, all three women coincidentally share a different approach towards disclosing and defending their childfree status to their parents-in-law.

Aisha, Nidhi and Meera do not have close relationships with their parents-in-law. Instead, each woman asks her partner to communicate with them and manage their expectations. Aisha has cut ties with her in-laws for reasons unrelated to being childfree. However, her husband is in touch with his parents and has informed them that they are not ready to have a child. This is a lie, a coping strategy to stave off the pressure to have children. Nidhi's mother-in-law has pressured Nidhi and her husband to change their stance. Nidhi is not bothered by this. She says, 'I don't say much to them because they're not my parents. I feel like it is his job to handle that area, and mine is to handle my side of the family.' Having children can close the distance between mothers and their parents-in-law because it creates another family connection – parents-in-law become the grandparents of the woman's child. Conversely, being childfree allows the women in this study to maintain some distance and protect themselves from stigmatization by their in-laws. However, this is only possible since the women live apart from their partner's parents and are not dependent on them.

Stigma Management and Coping Strategies by Childfree Women

Being childfree is an invisible, stigmatized attribute. This means that childfree individuals are not treated differently until they disclose that they are childfree. Stigmatization is characterized by exclusion, including negative stereotyped beliefs, prejudiced attitudes and discriminatory behaviour (Fudge Schormans, 2014: 6337). In response, stigmatized individuals, such as childfree women, can adopt stigma management and coping strategies based on the disclosure or non-disclosure of their childfree status. There are broadly two approaches to stigma management that childfree women can adopt: they can either disclose their status and manage the incoming response or hide their status altogether to avoid stigmatization. The study participants claimed that, rather than exclusionary behaviours, the stigmatization they experienced was primarily verbal.

The most common form of verbal stigmatization faced is microaggressions. Microaggressions are daily verbal, behavioural and environmental indignities, whether intentional or unintentional, that communicate hostile, derogatory, or negative slights and insults to marginalized persons or groups (Sue 2010: 5). Unintentional micro-aggressions reveal the stronghold of social normative discourse. The childfree women interviewed in the study have heard micro-aggressive statements from family members, friends, colleagues, partners and strangers. The younger women are not taken seriously, and the older women are criticized for their childfree position. The statements tend to be dismissive and condescending. These are some of the statements the participants recalled hearing:

1. Everyone has to do it [have children].
2. You will change your mind in the future.
3. You're just talking bullshit; you will have to have them [children].
4. Your partner will leave you for someone else.

The interviewees typically respond to micro-aggressions in two ways. They can ignore the comment by listening to it and not arguing back, thus ending the conversation. Otherwise, they can explain or justify their childfree position, extending the micro-aggression into a discussion or debate. The choice of method largely depends on whom the comment comes from. If the comment comes from colleagues or strangers, the women are mostly willing to let go of it without creating a fuss. If the comment comes from a family member or partner, the women are more likely to discuss it further. With more intimate and personal relationships, the women value the opinion and acceptance of the other party and are willing to extend the conversation. However, the participants insist they would not change their decision to be childfree, even if romantic partners or family members did not receive it well.

Questions and debates are arguably another form of verbal stigmatization as they often carry veiled prejudice and judgement within them. The most frequently heard response is, 'Who will take care of you when you are older?' Within this question lies a hidden implication that you need a child to take care of you in your old age. Divya asserts that a childfree person can care for themselves with proper financial management.

> All that money that would be spent on your child's tuition—you put all that money in your bank, and then you can finance your retirement very easily. If no husband is with me, if nobody is with me, this is what I'll do. Otherwise, who cares? Your own children leave you. Your own children die; what will you do then? You'll run into this problem [of taking care of yourself in old age] even if you have a child.

Nidhi shares the radical solution she and her partner prefer for themselves in their later years.

> I would rather go to Switzerland and euthanise myself than go through all this bullshit [relying on others for later-life care]. This is my old age plan. When I feel like I am satisfied with whatever I've done, my partner and I plan on going to dignitas.org and euthanising ourselves together. Nobody asked us if we wanted to be born, but at least we will have complete control over how and when we die.

Arguing back is a form of stigma management. You can challenge arguments by questioning the veiled normative ideology within them and denying the stigma. Divya and Nidhi do so by pointing out logical fallacies and sharing anecdotal counter-evidence (Meisenbach 2010: 278). Challenging micro-aggressions and condescending questions can also present an opportunity for education and destigmatization. Nidhi shared how she had to explain the term 'childfree' to those around her and how she had to combat misinformation and prejudice when talking to arranged marriage suitors.

> Very rarely did people question my choice. However, some guys I met through the arranged marriage process argued back with me. I once had a heated argument where a man said, 'That's a weird thing—that you only want to have sex, but you don't want to have kids, right?' I had to reply that [the two things] are not really related.

> Even people who have kids continue to have sex without having more kids. Because if you don't use contraception, I think you'll end up with 20 or 27 kids. Or the women die. [The man] was too dumb. I would just laugh at it. It didn't trigger me in any way. I was just like, okay, this person is just stupid, that's all.

These strategies are meant to handle stigmatic responses after disclosing their childfree status. However, there is another set of strategies that can be used to avoid disclosure and prevent any backlash. The interviewees commonly used these strategies to avoid telling their potential or current mothers-in-law the truth. If the childfree woman reveals her position, her partner's mother might object to the relationship. If they are already married, the couple needs a strategy to cope with mounting pressure to have children. The possible objections and applied pressure are forms of stigmatization that require strategies to be adequately managed. A couple of these strategies were discussed earlier.

1. Identity substitution – Divya's ex-boyfriend was willing to pretend to be infertile to avoid telling his mother that they were not planning to have children.
2. Fabrication – Aisha's husband staves off his parents' pressure to have children by lying, saying that the couple is not yet ready.

The participants outlined different stigma management and coping strategies they resort to when faced with actual or potential stigmatization. Depending on the person they were in a relationship with or interacting with, they chose whether or not to disclose being childfree and the appropriate corresponding strategy. However, the women stated that most of the time, they were open about being childfree, and their primary approach to dealing with microaggressions or critical questions was to disengage by not responding, changing the topic or cutting the conversation short, saying that it is okay to differ in opinion. All the participants said they are tired of the repeated, similar reactions they receive, and they wanted their childfree decision to be respected instead of challenged all the time.

Non-motherhood Forms of Caregiving

Though participants abstained from motherhood, they were engaged in other forms of caregiving, such as caring for siblings, parents, nieces or nephews, pets and others. Two of the participants – Anushka and Meera – have cared for pets. They shared that they significantly prefer pets over children. Anushka shared that she has cared for her pets more than her nieces and younger cousins. She has cared for cats, dogs, parrots, and many other pets over her lifetime. Like Anushka, Meera is passionate about animals and feels 'maternal' towards her two dogs. She doubts children can give her the 'same kind of love' her dogs do.

Nidhi vaguely talked about caring for her mother emotionally from a young age, which ultimately led to her decision to be childfree.

> I have something called 'Mommy Wound' [Mother Wound], which causes me to look for a mother figure a lot. It is a slang term that we use in psychology, referring to a situation where you were parentified by your parents. You had to take responsibility for your parents' feelings. You are wounded as a child, and then you feel like you have to assume that responsibility and become their parent. So, the child becomes parentified. As they grow up, they miss out on their childhood, and then they want to live their childhood as adults.

Being a mother is not the only form of caregiving. These interviews revealed that care may be needed and provided in other relationships, such as with pets or family relatives. The pet parents, Anushka and Meera, did not see pet parenthood as a replacement for motherhood but rather as preferable and better than motherhood. Caregiving can be difficult as well. Nidhi experienced a complicated childhood relationship with her mother, and her choice to be childfree now can be viewed as a form of finally caring for herself. These nuanced narratives of care directly counter stigmatic assumptions placed on childfree women in India.

Key Insights: Childfree Women's Narratives in Urban India

This study took an in-depth look into the lives of five childfree women living in urban India, and these were the key insights:

1. Every participant identified primary and secondary reasons for being childfree. Reasons included fear of pregnancy, lack of desire for motherhood, personal pursuits, career ambitions and trauma-related factors.
2. Every participant shows different degrees of openness or willingness to disclose their childfree status, depending on the person they are engaging with. They are more open with romantic partners and family members than with colleagues and friends.
3. In romantic relationships, early disclosure and similar childfree perspectives are preferred to ensure alignment in values and long-term goals.
4. In workplace environments, differences in childbearing preferences do not adversely impact colleague relationships.
5. Differences in childbearing preferences may cause strain in friendships as individuals grow older and drift apart when their life paths diverge and their priorities differ. Childfree individuals may then turn to social media to seek friendship and support.
6. The childfree participants were open about their position to their parents, receiving mixed responses ranging from dismissal to acceptance.

7. The childfree status of the women was often hidden from the partner's mother to protect the relationship and prevent stigmatization of the women.
8. There is almost a reflexive positive relationship between career prospects and being childfree. Unwillingness to compromise on career ambitions is one of the reasons the participants forgo motherhood. At the same time, not having children and being childfree boost career aspirations due to greater freedom of choice, higher mobility and other factors.
9. The most common type of stigmatization the participants experienced was verbal, in the form of micro-aggressions and questioning. The choice of stigma management strategy depended on whether the participant wanted to disclose their childfree status.
10. Post-disclosure stigma management strategies adopted by participants included ignoring comments, justifying their childfree status, challenging veiled assumptions through logical arguments, educating others on childfree lifestyles and normalizing the childfree position.
11. To avoid stigmatization by concealing childfree status, participants and their partners employed strategies like identity substitution, fabrication and passing.
12. Childfree women may forgo motherhood, but they do not forsake care. They might engage in other forms of caregiving besides being mothers, such as caring for pets and other family members.

Traces of the patriarchal ideologies behind the institution of motherhood in India are evident in the stigmatization that the childfree urban women in this study experience within their relationships and in their daily lives. The stigmatization does not affect the women enough to change their childfree stance, though. Most of the reasons behind being childfree uncovered in this study intersect with those detailed in the research works of Nandy (2013;

2014; 2017) and Bhambhani and Inbanathan (2018). The latter duo discussed the rational basis of decision-making among childfree individuals, which was also observed in this study. However, this qualitative study expanded further upon the decision to be childfree by identifying primary and secondary reasons and detailing the decision-making process and timelines. The stigma management strategies adopted by the childfree women interviewed are essentially the same as those discussed in Goffman's theory of stigma (1963) and childfree studies based on it (Park 2002; Verniers 2020). The strategies might be the same, but they are deployed uniquely in the Indian context.

Existing scholarly research on childfree women, especially in India, has not adequately addressed micro-level aspects such as the impact of being childfree on interpersonal relationships, daily interactions and personal outlooks. This chapter has made inroads into those unexplored areas of research and has furthered the study of stigma management strategies that childfree women adopt within Indian social settings.

Conclusion

Childfree women in India are often labelled lazy, selfish and irresponsible (Nandy 2013: 56), but the insights in this chapter challenge these assumptions. The study revealed that childfree women are ambitious and take control over their fertility preferences and life course. The time freed up by the lack of childrearing is occupied by personal and professional pursuits. Childfree women act in self-interest, but it is a mischaracterization to call them selfish. They care for others, such as family members, romantic partners, pets and others. They also demonstrate care and concern towards their hypothetical children. They would rather forgo motherhood than be an unfit or less-than-ideal mother to a child. Choosing to be childfree is considered a positive choice and an essential step towards looking out for themselves within the patriarchal traditionalist structures of Indian society. Their responsibility is also evidenced by their careful decision to be childfree, a rational one that is drawn out over time with critical evaluation and deep reflection.

Childfree people find themselves a marginalized minority in Indian society that has been neglected, criticized and ridiculed. By defying social expectations, childfree women invite stigmatization upon themselves for straying from tradition. However, despite the stigmatization, childfree women do not step away from their positions and beliefs. Instead, the number of childfree women is steadily growing in India, especially in urban areas. This is because women are empowered by the neoliberal modern economy, allowing them to lead independent lives apart from men or family if necessary. Self-determination proves itself stronger than stigmatization. Though stigma persists, childfree women navigate it while pursuing their goals and ambitions. However, stigma remains a hindrance that hampers well-being and prevents childfree women from achieving their full potential. Studying the micro-aspects of stigma management will contribute towards understanding the broader childfree phenomenon and shed light upon the macro demographic and socio-economic changes that will emerge shortly in India.

References

Basten, Stuart. 2009. 'Voluntary Childlessness and Being Childfree'. In *The Future of Human Reproduction: Working Paper #5*, 9. Oxford & Vienna: St. John's College, Oxford & Vienna Institute of Demography.

Berrington, Ann. 2017. 'Childlessness in Europe: Contexts, Causes, and Consequences'. In *Demographic Research Monographs,* edited by M. Kreyenfeld and D. Konietzka, 58. Cham: Springer International Publishing.

Bhambhani, Chandni and Anand Inbanathan. 2018. 'Not a Mother, yet a Woman: Exploring Experiences of Women Opting out of Motherhood in India'. *Asian Journal of Women's Studies* 24(2): 159–82.

Fudge Schormans, Ann. 2014. 'Stigmatization'. In *Encyclopedia of Quality of Life and Well-Being Research*, edited by A. C. Michalos, 6336–41. Dordrecht: Springer.

Goffman, Erving. 1963. *Stigma: Notes on the Management of Spoiled Identity*. New York: Simon & Schuster, 3–73.

Gupta, Neha. 2017. 'Being Childless by Choice in the Indian Society'. *Times of India*, 16 May. Available online: https://timesofindia.indiatimes.com/life-style/relationships/soul-curry/being-childless-by-choice-in-the-indian-society/articleshow/58695326.cms (accessed on 13 May 2025).

International Institute for Population Sciences (IIPS) and ICF 2021. 'National Family Health Survey (NFHS-5), 2019–21: India: Volume I'. In *National Family Health Survey, India*. Mumbai: IIPS. Available online: https://dhsprogram.com/pubs/pdf/FR375/FR375.pdf (accessed on 3 June 2025).

Meisenbach, Rebecca J. 2010. 'Stigma Management Communication: A Theory and Agenda for Applied Research on How Individuals Manage Moments of Stigmatized Identity'. *Journal of Applied Communication Research* 38(3): 278.

Nair, Eisha. 2022. 'As Indian Couples Increasingly Choose to Go Childfree, Premarital Counselling Must Be Extended to Their Families'. *Firstpost*, 9 January. Available online: https://www.firstpost.com/art-and-culture/as-indian-couples-increasingly-choose-to-go-childfree-premarital-counselling-must-be-extended-to-their-families-10263391.html (accessed on 1 June 2025).

Nandy, Amrita. 2013. 'Outliers of Motherhood: Incomplete Women or Fuller Humans?' *Economic and Political Weekly* 48(44): 53–9.

Nandy, Amrita. 2014. 'Shadow of Motherhood: Writing the Outlier Self'. *UCLA Thinking Gender Papers*. Available online: https://escholarship.org/uc/item/38x7122z (accessed on 4 May 2025).

Nandy, Amrita. 2017. *Motherhood and Choice: Uncommon Mothers, Childfree Women*. New Delhi: Zubaan, 63.

Park, Kristin. 2002. 'Stigma Management among the Voluntarily Childless'. *Sociological Perspectives* 45(1): 21–45.

Singh, Raghuvendra. 2019. 'Vidya Balan Breaks Her Silence on Rumours about Her Being Pregnant'. *Filmfare*, 12 August. Available online: https://www.filmfare.com/news/bollywood/vidya-balan-breaks-her-silence-on-rumours-about-her-being-pregnant-35487.html (accessed on 15 May 2025).

Sue, Derald Wing. 2010. *Microaggressions in Everyday Life: Race, Gender, and Sexual Orientation*. New Jersey: John Wiley & Sons, 5.

Suri, Rishabh. 2019. 'Akshaye Khanna on Not Tying the Knot or Adopting Kids: I'm Not marriage Material, Can't Give Up Control'. *Hindustan Times*, 18 September. Available online: https://www.hindustantimes.com/bollywood/akshaye-khanna-on-not-tying-the-knot-or-adopting-kids-i-m-not-marriage-material-can-t-give-up-control/story-YlBl8iGWEIc0y4Mm0XUVoN.html (accessed on 11 May 2025).

United Nations, Department of Economic and Social Affairs, Population Division 2024. 'World Population Prospects: The 2022 Revision'. Available at: https://population.un.org/wpp/ (accessed on 7 June 2025).

Verniers, Catherine. 2020. 'Behind the Maternal Wall: The Hidden Backlash toward Childfree Working Women'. *Journal of Theoretical Social Psychology* 4(3): 107–24.

About the Editor and Contributors

About the Editor

Jagriti Gangopadhyay is Associate Professor of Sociology at the Manipal Institute of Social Sciences Humanities and Arts, Manipal Academy of Higher Education, Manipal, India. Her research lies at the intersections of social gerontology and public health. She has published widely in peer-reviewed journals and edited volumes and contributes to both scholarly and public discourse on ageing practices and public health issues. She was awarded the 'Shastri Publication Grant' by the Shastri Indo-Canadian Institute for her monograph titled *Culture, Context and Aging of Older Indians: Narratives from India and Beyond* (2021), published by Springer, Singapore. She has also co-edited the book *Eldercare Issues in China and India* (2022), published by Routledge, UK.

About the Contributors

Aashna Viswanathan is a PhD scholar at the Manipal Institute of Social Sciences Humanities and Arts, Manipal Academy of Higher Education, Manipal, India. She is currently completing her doctoral dissertation on gender, sexuality and literary form in contemporary retellings of the Indian Epics. Her research studies the complexities around representations of gender, sexualities, bodies and desires in post-millennial retellings of the Indian Epics. Her academic interests lie in gender and sexuality studies, post-millennial literature and popular culture.

Anubhav Sengupta is Associate Professor of Sociology at the Manipal Institute of Social Sciences Humanities and Arts, Manipal

Academy of Higher Education, Manipal, India. He holds a Doctor of Philosophy (PhD) degree in Sociology from Jawaharlal Nehru University, New Delhi, India. For the completion and submission of his thesis, he was awarded the SYLFF fellowship for the period 2013–16. His scholarly interests encompass student and youth movements in India, radical movements, subjectivity studies, political sociology and political theories.

Anusha Ravishankar is a PhD scholar at the Manipal Institute of Social Sciences Humanities and Arts, Manipal Academy of Higher Education, Manipal, India, and Curator of the Hebbar Gallery and Art Centre in Manipal, India. Her doctoral research project focuses on alternative performance spaces in urban India. Her multidisciplinary background includes the curation of festivals, publications and spaces; research on performing arts theory and archives; and advocacy for theatre in school curricula. She has worked with leading institutions in India, such as the National Museum, Indira Gandhi National Centre for the Arts, Partition Museum and Ranga Shankara Theatre. Her artistic practice explores intersections in performance installations and movement. Her solo and collaborative works have been presented to audiences in India, the UK and Serbia.

Beniamin Kłaniecki is Assistant Professor at the Faculty of English, Adam Mickiewicz University, Poznań, Poland. He earned his PhD in Postcolonial Literature in 2021. His research lies at the intersection of literature, queer studies and masculinity studies, with a primary focus on post-millennial Indian fiction in English. He is currently working on representations of human corporeality in post-millennial Polish and Indian English-language speculative fiction, as well as literary discourses on masculinity in the context of Gulf migration. His work has been published in the *Journal of Postcolonial Writing*, *Critique: Studies in Contemporary Fiction* and *Literature, Critique and Empire Today*.

Carmen Wong is an artist-researcher and creative educator with extensive experience in the arts, community, food systems and higher education. Dedicated to food sovereignty, racial equity and social justice, particularly in the fields of arts, performance and intersections in food communities, she is growing skills rooted in non-violent communication and deep listening. Carmen is a co-founder and co-animator of JarSquad, a social art project based in Plymouth, UK, growing a hyperlocal solidarity economy with communal preserves made using food surplus.

Gayathri Prabhu is Professor of Literary Studies and the Incharge Director at the Manipal Institute of Social Sciences Humanities and Arts, Manipal Academy of Higher Education, Manipal, India. She is the author of eight books, most recently *A Genre of Her Own: Life Narratives and Feminist Literary Beginnings in Modern India* (Bloomsbury Academic, 2025).

Jayaditya Vittal is a PhD scholar at the Manipal Institute of Social Sciences Humanities and Arts, Manipal Academy of Higher Education, Manipal, India. His thesis examines narratives of place in Basroor, an estuary town in the foothills of the Western Ghats. His work revolves around ecocriticism, environmental humanities and narratives of place in the Western Ghats of India.

Melvin Mathew Thomas is a PhD scholar at the Manipal Institute of Social Sciences Humanities and Arts, Manipal Academy of Higher Education, Manipal, India. He is a researcher whose work centres at the intersection of student mental health and education, with the goal of shaping policies and interventions that foster inclusive and empathetic learning environments. His academic interests include film theory and criticism, visual anthropology, disability studies, medical humanities and media theory. With a background in teaching, he has facilitated courses on these subjects for both undergraduate and postgraduate students. Through his interdisciplinary approach, Melvin aims to bridge critical inquiry with evolving practice.

Nikhil Govind is Professor at the Manipal Institute of Social Sciences Humanities and Arts, Manipal Academy of Higher Education, Manipal, India. He is the author of *Between Love and Freedom: The Revolutionary in the Hindi Novel* (2014), Routledge, UK; *Inlays of Subjectivity: Affect and Action in Modern Indian Literature* (2019), Oxford University Press India; *Shadow Craft: Visual Aesthetics of Black and White Hindi Cinema* (with Gayathri Prabhu, 2020), Bloomsbury India; and *The Moral Imagination of the Mahabharata* (2022), Bloomsbury India.

Rhea Jaikumar Menon is a PhD scholar at the Manipal Institute of Social Sciences Humanities and Arts, Manipal Academy of Higher Education, Manipal, India. Her research interests span various subfields of sociology, including social demography and medical sociology. She investigates the implications of emerging demographic trends and patterns, and her doctoral research is centred on the lives of childfree individuals in urban India. She has previously published articles in the journals *Anthropology & Aging* and *Society* and contributed a chapter to the *Handbook of Aging, Health and Public Policy: Perspectives from Asia*, published by Springer Singapore.

Shreyas Kolpe is a PhD scholar at the Manipal Institute of Social Sciences Humanities and Arts, Manipal Academy of Higher Education, Manipal, India. He is a researcher on the history of caste in coastal Karnataka. His other academic interests are regional history and historiography, epigraphy and the history of religion. With training in computer science, he is also interested in aspects of digital humanities.

Sre Ratha is a PhD scholar at the Manipal Institute of Social Sciences Humanities and Arts, Manipal Academy of Higher Education,

Manipal, India. Her doctoral work focuses on the everyday problems of Indian mothers in STEM academia. Her research interests include gender and violence, gender-based policy and the feminism of science. She has published with the *Indian Journal of Medical Ethics* and has presented her work at national and international conferences.

Vishvajith Madhavamurthy is a PhD scholar at the Manipal Institute of Social Sciences Humanities and Arts, Manipal Academy of Higher Education, Manipal, India. His work explores the intersections of everyday activism, nationalist ideologies and sub-national identities, with a focus on shaping discourse around protectionist politics. His academic interests extend to language politics, identity politics, nationalism, neoliberalism, and the relationship between society and cinema. Combining film analysis with field interviews, his MA research examined the construction of the ideal Tuluva identity in Tulu and Kannada films, comparing cinematic narratives with lived perspectives among the Tuluva people. Through his interdisciplinary approach, Vishvajith seeks to bridge critical inquiry with evolving socio-political realities.

Index